Springer Theses

Recognizing Outstanding Ph.D. Research

Aims and Scope

The series "Springer Theses" brings together a selection of the very best Ph.D. theses from around the world and across the physical sciences. Nominated and endorsed by two recognized specialists, each published volume has been selected for its scientific excellence and the high impact of its contents for the pertinent field of research. For greater accessibility to non-specialists, the published versions include an extended introduction, as well as a foreword by the student's supervisor explaining the special relevance of the work for the field. As a whole, the series will provide a valuable resource both for newcomers to the research fields described, and for other scientists seeking detailed background information on special questions. Finally, it provides an accredited documentation of the valuable contributions made by today's younger generation of scientists.

Theses are accepted into the series by invited nomination only and must fulfill all of the following criteria

- They must be written in good English.
- The topic should fall within the confines of Chemistry, Physics, Earth Sciences, Engineering and related interdisciplinary fields such as Materials, Nanoscience, Chemical Engineering, Complex Systems and Biophysics.
- The work reported in the thesis must represent a significant scientific advance.
- If the thesis includes previously published material, permission to reproduce this must be gained from the respective copyright holder.
- They must have been examined and passed during the 12 months prior to nomination.
- Each thesis should include a foreword by the supervisor outlining the significance of its content.
- The theses should have a clearly defined structure including an introduction accessible to scientists not expert in that particular field.

More information about this series at http://www.springer.com/series/8790

Ryuji Takahashi

Topological States on Interfaces Protected by Symmetry

Doctoral Thesis accepted by
Tokyo Institute of Technology, Tokyo, Japan

Springer

Author
Dr. Ryuji Takahashi
Department of Physics
Tokyo Institute of Technology
Tokyo
Japan

Supervisor
Prof. Shuichi Murakami
Department of Physics
Tokyo Institute of Technology
Tokyo
Japan

ISSN 2190-5053 ISSN 2190-5061 (electronic)
Springer Theses
ISBN 978-4-431-55533-9 ISBN 978-4-431-55534-6 (eBook)
DOI 10.1007/978-4-431-55534-6

Library of Congress Control Number: 2015934436

Springer Tokyo Heidelberg New York Dordrecht London

Printed on acid-free paper

Springer Japan KK is part of Springer Science+Business Media (www.springer.com)

Parts of this thesis have been published in the following journal articles:

- R. Takahashi and S. Murakami, "Interfacial Fermi Loops from Interfacial Symmetries", Phys. Rev. Lett., **113**, 256406 (2014).
- R. Takahashi and S. Murakami, "Completely Flat Bands and Fully Localized States on Surfaces of Anisotropic Diamond-Lattice Models", Phys. Rev. B, **88**, 235303 (2013).
- R. Takahashi and S. Murakami, "Gapless Interface States between Topological Insulators with Opposite Dirac Velocities", Phys. Rev. Lett., **107**, 166805 (2011).

Supervisor's Foreword

Interfaces between two solids often exhibit exotic phenomena that are absent in either of the two parental materials. Interfaces have a number of degrees of freedom such as a combination of two materials, crystallographic orientations of the interface, and even the quality of the interface. Interface electronic states are often quite sensitive to these factors. This sensitivity to a number of factors, some of which cannot be controlled easily, causes various difficulties in theoretical research on interface electronic states and comparison with experiments. Theories on electronic states on interfaces are often restricted to first-principle calculations and phenomenological theories, because simplified effective models may not capture basic properties of interface states.

Historically, surface physics had had a similar aspect. Nevertheless, the discovery of topological insulators has added a renewed interest in the surface physics. It is qualitatively new that some surface electronic properties are robust and are determined by bulk topological properties. Furthermore, such proposals have been confirmed experimentally in various materials.

This book is concerned with theoretical pursuit of such topological phenomena in interface physics. This book, based on the dissertation of Dr. Ryuji Takahashi, theoretically explores possibilities of novel interface states from an interplay between topology and symmetry. Because the topics in this book are based on topology and symmetry, even simplified models with required properties of topology and symmetry turn out to be powerful in predicting new interface phenomena, which is in contrast with conventional interface phenomena. Various predictions in this book are yet to be observed experimentally, but the abovementioned varieties of interfaces suggest that there is much room for future experimental and theoretical investigations.

Tokyo, Japan
January 2015

Prof. Shuichi Murakami

Acknowledgments

I would like to thank my Ph.D. advisor, Professor Shuichi Murakami, for his patient supervision. I gained knowledge from him about various subjects in condensed matter physics. He taught me how to tackle research. Without his support, I could not have accomplished this work. I am also very grateful for his support for my travels and participation in many conferences and workshops.

I would also like to thank the faculty members, secretaries, and colleagues in the theoretical condensed matter physics groups in the Department of Physics at the Tokyo Institute of Technology, and Dr. Daniel Gosálbez-Martínez in the Universidad de Alicante in Spain.

Among Assistant Professors, I would especially like to thank Dr. Takehiko Yokoyama, Dr. Ryuichi Shindou, and Dr. Kiyomi Okamoto. Dr. Yokoyama gave me critical comments about my research. Through his work, I became interested in the finite-size effect in topological insulators. Dr. Shindou has great knowledge of many fields in physics. From him, I learned approaches for solving problems in physics. Dr. Okamoto and I worked in the same office for 5 years, and sometimes I had helpful discussions with him.

This work has been supported by the Global Center of Excellence Program of the Ministry of Education, Culture, Sports, Science, and Technology of Japan ("MEXT"), through the "Nanoscience and Quantum Physics" Project of the Tokyo Institute of Technology.

Finally, I would like to express my deep appreciation to my family, including my cousins.

Acknowledgments

[illegible]

Contents

1 Introduction 1
1.1 Topological Phase of Matters 1
1.2 Organization of the Dissertation 3
References 4

2 Topological Invariant and Topological Phases 5
2.1 Integer Quantum Hall Effect 5
2.1.1 Integer Quantum Hall Effect as a Topological Insulator 7
2.1.2 Adiabatic Charge Polarization by the Berry Phase 9
2.1.3 Laughlin's Gedanen Experiment 11
2.1.4 Physical Picture of the Edge State 14
2.1.5 Berry Curvature in Systems with Broken Time-Reversal Symmetry 16
2.2 Topological Insulator 16
2.2.1 Topological Invariant Under Time-Reversal Symmetry . . . 18
2.2.2 Surface Effective Hamiltonian 24
2.3 Topological Flat-Band States in Honeycomb Lattice 25
2.3.1 Dispersion of the Honeycomb Lattice Model 25
2.3.2 Topological Explanation for Existence of the Flat-Band States 29
2.3.3 Completely Localized Edge States 30
References 33

3 Gapless Interface States Between Two Topological Insulators 35
3.1 Refractive Phenomena of the Surface States at the Boundary of Two Topological Insulators 35
3.2 Gapless Interface States Between Two TIs 39
3.2.1 Paradox in the Refractive Phenomena 39
3.2.2 Gapless States from the Surface Effective Model 40

3.2.3 Proof for the Existence of the Gapless States by the Mirror Chern Number 43
3.3 Gapless Interface States in a Lattice Model 46
3.3.1 Model 46
3.3.2 Calculation of the Mirror Chern Number for the Model 47
3.3.3 Gapless Interface States 53
3.4 Interfacial Fermi Loops and Interfacial Symmetries 55
3.4.1 Interfacial Fermi Loops in the FKM Model. 55
3.4.2 Interfacial Fermi Loops from Interfacial Symmetries 59
3.4.3 Interfacial Fermi Loops in the π-junction Interface. 61
References 62

4 Weyl Semimetal in a Thin Topological Insulator 63
4.1 Gapless States Protected by Mirror Symmetry in a Topological Insulator 63
4.1.1 Weyl Semimetal Under Broken Inversion Symmetry 67
4.2 Weyl Semimetal Phase in a Lattice Model 68
4.3 Estimation of the Magnetization at the Phase Transition 70
References 71

5 Summary and Outlook 73

6 Properties of the Chern Numbers 77
6.1 Chern Number. 77
6.1.1 Berry Curvature with Time-Reversal Symmetry. 78
6.2 Kramers Theorem 79
6.3 Berry Curvature and Chern Number. 80
6.3.1 Change of the Chern Number 81
Reference 83

7 Calculation for the Interface Fermi Loops. 85
7.1 Classification of Interfacial Systems 85
7.2 Fermi Loop for the IPHS 86
Reference 88

Curriculum Vitae 89

Chapter 1
Introduction

1.1 Topological Phase of Matters

In a two-dimensional electron gas subjected to a strong magnetic field at low temperature, the energy spectra take on discrete values, which are known as Landau levels. When the Fermi energy is in a gap between Landau levels, the system becomes insulating, while the Hall conductance is measured as quantized values in units of e^2/h. The phenomenon, called the quantum Hall effect, was discovered experimentally [1, 3, 4].[1] Intriguingly, the value of the unit of quantization is closely related with a universal quantity in physics, which is well known as the fine-structure constant in quantum electrodynamics: $\alpha = \frac{e^2}{c\hbar} \sim \frac{1}{137}$. The Hall conductance does not depend on details of the system, such as impurities or its geometry, as long as the gap between the Landau levels is sufficiently large. Moreover, researches on the quantum Hall effect found out novel states which are localized on the edge of the system. In the quantum Hall effect, the Hall current is carried by the edge states. If an impurity is on the edge, the edge current goes around it and perfectly transmits without backscatterings. This is because the edge current flows in one-way; there are no states going in the opposite direction of the velocity along the edge. This one-way propagation is called chiral. This chiral edge mode is attributed to the cyclotron motion. Because these boundary states are originating from topological order of the bulk, the edge current is robust against deformation of the edge or weak perturbations. This is the first example of the bulk-boundary correspondence.

As a related subject, topological insulators were found in time-reversal invariant systems, which were proposed theoretically [5–9]. The bulk of these materials are insulating while there are metallic states at the boundary. The effective model of such boundary states is described by the massless Dirac equation, and the dispersion of the boundary states across the bulk gap typically forms a cone, called the Dirac cone. The electrons in the system show spin-filtered phenomena caused by the

[1] Before the experimental discovery [1], Ando et al. reported that the Hall conductivity is quantized in a two dimensional system subjected to a magnetic field [2].

R. Takahashi, *Topological States on Interfaces Protected by Symmetry*,
Springer Theses, DOI 10.1007/978-4-431-55534-6_1

spin-orbit interaction; the electrons running in opposite directions have opposite spins on the boundary. This property of the boundary mode is called helical. and it prohibits perfect backscattering by nonmagnetic impurities. A two-dimensional topological insulator is also called the quantum spin Hall system, and similarly to the quantum Hall effect, the conductance is quantized as $2e^2/h$ carried by the edge states. This helical edge mode transmits perfectly as long as the system retains time-reversal symmetry. The quantized Hall conductance was observed experimentally in HgTe quantum well [10], quickly after the theory was proposed [7]. Unlike the quantum Hall effect, the spin-filtered boundary state is also realized in three-dimensional systems. Three-dimensional time-reversal invariant insulators are topologically classified into two phases: the weak and strong topological insulators. In weak topological insulators, there are an even number of Dirac cones as the surface states, which are weak against perturbations such as impurities; therefore in this respect we usually regard the system to be similar to an ordinary insulator. On the other hand, in strong topological insulators, there are an odd number of Dirac cones as the surface states. They are robust against perturbations because the gapless states are guaranteed by the Kramers theorem, and when we call the topological insulator, it generally means the strong topological insulator. They are observed by the angle-resolved photoemission spectroscopy [11–14].

Generally, topological orders depend on symmetries; time-reversal symmetry is broken in quantum Hall systems, while in topological insulators time-reversal symmetry is preserved. In addition, the gap plays an important role in the topological protection. In band insulators, the conduction and valence bands are separated by the gap. Therefore, against external perturbations such as impurities, the topological order remains intact due to the gap. Namely, the system is not topologically affected by weak perturbations compared with the bulk gap. The definition of the topological equivalence in insulators is given as follows [15],

Definition 1.1 (*Topological equivalence*) Band structures are equivalent if they can be continuously deformed into each other without closing the energy gap and with conservation of the symmetries.

In addition to the deformation, if topologically trivial bands are added in the band structure, the system is topologically the same as before. This notion of topological equivalence leads to the definition of the topological number, e.g. the Chern number in quantum Hall effect, and the Z_2 topological number in topological insulators. The topological number labels topologically ordered phases, which are classified by topological equivalence.

Besides the topological classes for the cases with or without time-reversal symmetry, the topological phases protected by spatial symmetry have been studied [16, 17]. These researches shed light on the weak topological insulators, and led to discoveries of new classes of topological metallic surface states. In general, the gapless surface states in weak topological insulators are not robust against external effects. These kind of gapless states are not characterized by the Z_2 topological number but other topological numbers based on spatial symmetries as a source of the topological protection. We will discuss topological phenomena which are based on this notion in Chaps. 3 and 4.

The above-mentioned phenomena are originating from single-particle non-interacting physics. The difference among these topological classes is whether or not the time-reversal symmetry is broken. In systems without time-reversal symmetry, an additional topological phenomena arise due to electron-electron interaction [4], known as the fractional quantum Hall effect. Just as the name indicates, the phenomenon is caused by fractional Landau-level fillings, and the conductivity is also fractional in units of e^2/h. The fractional Hall effect shows rich physics caused by the fractional statistics of the anyon, and has been studied intensively. As is similar to the quantum Hall effect, research interests in topological insulators are spreading to correlated systems to realize the fractional topological insulators. However, there are various physical phenomena by using topological insulators even within single-particle physics. In this dissertation, we will discuss several topics based on single-particle physics. Topological insulators are mainly used in achieving the physical phenomena discussed in the dissertation.

1.2 Organization of the Dissertation

In this dissertation, we theoretically study various aspects of topological states on surfaces and interfaces. For physical quantities and in processes of the calculation, the following physical constants are frequently expressed to be unity: the velocity of light c, the Plank constant $\hbar$, the charge of an electron, and the lattice constant. When physical quantities are expressed by physical constants, we explicitly write the fundamental constants listed above. In equations, when we express differentials in k-space, we frequently omit k without explanation, e.g., $\frac{\partial}{\partial k_x} = \partial_x$.

The organization of the dissertation is the following. In Chap. 2 we introduce topological states characterized by the Berry phase. We first introduce the integer quantum Hall effect, and show that the Hall conductivity is quantized by calculating the Berry phase. In addition, we show the quantized Hall conductivity by the edge picture. Then we give a brief introduction on topological insulators.

In Chap. 3 we first study refraction phenomena at a junction between two topological insulator surfaces with different velocities. We calculate reflection and transmission coefficients. Next we discuss the case when the velocities of surface states for the two topological insulators have different signs. In this case we show that gapless states appear on the interface and give a proof of the existence of the such gapless interface states by using the mirror Chern number. We also show the gapless states by a low-energy continuum model and the Fu-Kane-Mele tight-binding model. Furthermore, we show a novel dispersion of the interface states, i.e. the "Fermi loop", when the system has particle-hole symmetry by using the Fu-Kane-Mele tight-binding model. We give an explanation of the existence of the gapless states and make a formalism to characterize the Fermi loop in terms of interfacial symmetries.

In Chap. 4 we discuss the gapless nature of topological insulators subjected to an in-plane magnetic field in terms of mirror symmetry. In addition we study edge states in a Weyl semimetal realized in a thin topological insulator with an in-plane magnetization. We show that the edge states have a finite velocity when the inversion symmetry is broken.

In Chap. 5 we summarize this dissertation.

References

1. K. von Klitzing, K.G. Dorda, M. Pepper, Phys. Rev. Lett. **45**, 494 (1980)
2. T. Ando et al., J. Phys. Soc. Jpn. **39**, 279 (1975)
3. D.J. Thouless, M. Kohmoto, M.P. Nightingale, M. den Nijs, Phys. Rev. Lett. **49**, 405 (1982)
4. X.G. Wen, Adv. Phys. **44**, 405 (1995)
5. C.L. Kane, E.J. Mele, Phys. Rev. Lett. **95**, 226801 (2005)
6. C.L. Kane, E.J. Mele, Phys. Rev. Lett. **95**, 146802 (2005)
7. B.A. Bernevig, T.L. Hughes, S.-C. Zhang, Science **314**, 1757 (2006)
8. J.E. Moore, L. Balents, Phys. Rev. B **75**, 121306(R) (2007)
9. L. Fu, C.L. Kane, E.J. Mele, Phys. Rev. Lett. **98**, 106803 (2007)
10. M. König et al., Science **318**, 766 (2007)
11. D. Hsieh et al., Nature (London) **452**, 970 (2008)
12. D. Hsieh et al., Nature (London) **460**, 1101 (2009)
13. Y. Xia et al., Nat. Phys **5**, 398 (2009)
14. Y.L. Chen et al., Science **325**, 178 (2009)
15. M.Z. Hasan, C.L. Kane, Rev. Mod. Phys. **82**, 3045 (2010)
16. R. Takahashi, S. Murakami, Phys. Rev. Lett. **107**, 166805 (2011)
17. T.H. Hsieh et al., Nat. Commun. **3**, 982 (2012)

Chapter 2
Topological Invariant and Topological Phases

2.1 Integer Quantum Hall Effect

Historically in two-dimensional electron gas (2DEG) subjected to a strong magnetic field, the integer quantum Hall (QH) effect was observed experimentally for the first time by von Klitzing et al. [1]. In the QH effect, the Hall conductivity is given as $\nu\frac{e^2}{h}$ where ν is an integer. The quantized Hall conductivity was interpreted in terms of gauge invariance [2, 3] under flux insertion and later topological invariance was proposed [4] to interpret its physical origin. Here we explain the topological number ν given by Thouless, Kohmoto, Nightingale and den Nijs (TKNN) [4]. This ν number is called the TKNN number and was calculated from the Kubo formula at first. Later it was found to be interpreted as the Chern number given by an integral of the Berry curvature [5], as we explain in the following.

In quantum mechanics, when quantum states are subjected to an adiabatic transition, the Berry connection (phase) appears as a change of the phase of the wavefunction. In addition, the Hall current by the Kubo formula have the same form with this phase change. Besides the Kubo formula, the Hall current contributed from the Berry curvature can also be calculated from the Schrödinger equation as follows [6]. We consider a time-dependent 2D system with translational symmetry along the x and y axes. The Hamiltonian is written as $H(\mathbf{k}(t))$, where $\mathbf{k} = (k_x(t), k_y(t))$ is the wavevector, and the eigenstate and eigenvalue are $|u_{\mathbf{k}}\rangle$ and $E_{\mathbf{k}}$, respectively. The current along the x axis $I_x(\mathbf{k})$ is expressed as

$$\begin{aligned} I_x &\sim \left\langle u_{\mathbf{k}} \left| \frac{\partial H(\mathbf{k}(t))}{\partial k_x} \right| u_{\mathbf{k}} \right\rangle \\ &= \partial_x E_{\mathbf{k}} - \langle \partial_x u_{\mathbf{k}} | H(\mathbf{k}) | u_{\mathbf{k}} \rangle - \langle u_{\mathbf{k}} | H(\mathbf{k}) | \partial_x u_{\mathbf{k}} \rangle, \end{aligned} \tag{2.1}$$

where $\partial_x = \frac{\partial}{\partial k_x}$. Then by the Schrödinger equation,

$$H(\mathbf{k})|u_{\mathbf{k}}\rangle = i\frac{\partial}{\partial t}|u_{\mathbf{k}}\rangle = i(\dot{k}_x\partial_x + \dot{k}_y\partial_y)|u_{\mathbf{k}}\rangle, \tag{2.2}$$

R. Takahashi, *Topological States on Interfaces Protected by Symmetry*,
Springer Theses, DOI 10.1007/978-4-431-55534-6_2

we have

$$I_x(\mathbf{k}) \sim \partial_x E_{\mathbf{k}} - b_z(\mathbf{k})\dot{k}_y, \tag{2.3}$$
$$b_z(\mathbf{k}) = i(\langle \partial_x u_{\mathbf{k}} | \partial_y u_{\mathbf{k}} \rangle - \langle \partial_y u_{\mathbf{k}} | \partial_x u_{\mathbf{k}} \rangle). \tag{2.4}$$

The quantity $b_z(\mathbf{k})$ defined in Eq. (2.4) is called the Berry curvature. The first term $\partial_x E_{\mathbf{k}}$ in Eq. (2.3) corresponds to the velocity along the x axis. For the second term in Eq. (2.3), when an electric field $\mathbf{E}$ is applied along the y axis, the temporal differential of the wavenumber corresponds to the electronic field $\dot{k}_y \sim -\frac{eE_y}{\hbar}$, and the Hall conductivity is given as the Berry curvature $\sigma_{xy} \sim b_z(\mathbf{k})$. Therefore we have shown that the Berry curvature contributes to the Hall current. Furthermore,[1] when the system is gapped, the contribution from one of the occupied band to σ_{xy} is given by the sum of the Berry curvature over the Brillouin zone (BZ) as,

$$\sigma_{xy} = \nu \frac{e^2}{h}, \tag{2.5}$$
$$\nu = \frac{1}{2\pi} \int_{\mathrm{BZ}} \mathrm{d}^2 k b_z(\mathbf{k}), \tag{2.6}$$

where ν is an integer (see Sect. 6.1). This is the TKNN integer, also called the Chern number.

By extending the above discussion to general band structure, the Hall conductivity is expressed as

$$\sigma_{xy} = \nu \frac{e^2}{h}, \tag{2.7}$$
$$\begin{aligned}\nu &= \frac{i}{2\pi} \sum_{E_m < E_F} \int_{\mathrm{BZ}} \mathrm{d}^2 k (\langle \partial_x u_{m\mathbf{k}} | \partial_y u_{m\mathbf{k}} \rangle - \langle \partial_y u_{m\mathbf{k}} | \partial_x u_{m\mathbf{k}} \rangle) \\ &= \frac{1}{2\pi} \int_{\mathrm{BZ}} \mathrm{Tr}(\mathscr{F}),\end{aligned} \tag{2.8}$$

where the suffix m is the index of the bands from the lowest energy band, the summation is taken over the bands below the Fermi energy E_F, and

$$\mathscr{F} = \mathrm{d}a - ia \wedge a, \tag{2.9}$$

[1] We assume that the Berry curvature does not depend on time t, because the Hall current is the lowest order in the time-dependent term, when we assume that the time dependence of the system is weak.

with the matrix of the Berry connection $a_{mn} = i\langle u_{m\mathbf{k}}|\mathrm{d}u_{n\mathbf{k}}\rangle$.[2] The Chern number ν defined here is shown to be integer. Thus the quantization of the Hall conductivity follows for any band insulators without interaction and disorder.

2.1.1 Integer Quantum Hall Effect as a Topological Insulator

We have seen that the Hall current is described by the Berry curvature, and the Hall conductivity is quantized in the gapped system as $\nu\frac{e^2}{\hbar}$, where ν is an integer. Therefore, to observe the quantized Hall conductivity, ν must be nonzero. Here we discuss the origin of the Hall conductivity, in the case of an insulator with only one occupied band according to Kohmoto [7]. The Hall conductivity is written in terms of the connection as

$$\begin{aligned}\sigma_{xy} &= \frac{e^2}{\hbar}\int_{\mathrm{BZ}} \mathrm{d}\mathbf{k}[\nabla_{\mathbf{k}} \times \mathbf{a}]_z \\ &= \frac{e^2}{\hbar}\oint_{\mathrm{BZ}} \mathrm{d}\mathbf{k}\cdot\mathbf{a},\end{aligned} \tag{2.11}$$

where $\mathbf{a} = i\langle u_{n\mathbf{k}}|\nabla_{\mathbf{k}}u_{n\mathbf{k}}\rangle$, we omitted the band index for simplicity, and used the Stokes theorem. We assume that only the n-th band is occupied. The above expression means that the Hall conductivity is invariant under the gauge transformation,

$$|u_{\mathbf{k}}\rangle \to \mathrm{e}^{if_{\mathbf{k}}}|u_{\mathbf{k}}\rangle, \tag{2.12}$$

where $f_{\mathbf{k}}$ is real. On the other hand, the connection depends on the phase; the gauge transformation gives

$$\mathbf{a} \to \mathbf{a} + \nabla_{\mathbf{k}} f_{\mathbf{k}}. \tag{2.13}$$

In 2Ds, the BZ can be regarded as a torus, and the wavefunction is expected to be periodic. Naively, one might think that the above equation Eq. (2.11) vanishes due to the periodicity because the path of the integral contains the same states with opposite directions (see Fig. 2.1). However, ν becomes nonzero, when the phase of the wavefunction cannot be determined uniquely and smoothly in the whole BZ; in other words, the phase singularity in the BZ cannot be removed by any gauge transformation when ν is not trivial.

To investigate the relation between the phase and the Hall conductivity, we consider an arbitrary eigenstate $|u_{\mathbf{k}}\rangle$ on 2D systems, and we try to determine the phase of

---

[2] For a vector $\mathrm{x} = x_1\mathbf{e}_1 + x_2\mathbf{e}_2 + x_3\mathbf{e}_3$ where $\mathbf{e}_{i=1,2,3}$ are the basis vectors, the wedge product $\wedge$ is defined as

$$\mathrm{d}x_1 \wedge \mathrm{d}x_2 = \mathbf{e}_1 \times \mathbf{e}_2\mathrm{d}x_1\mathrm{d}x_2 = \mathbf{e}_3\mathrm{d}x_1\mathrm{d}x_2. \tag{2.10}$$

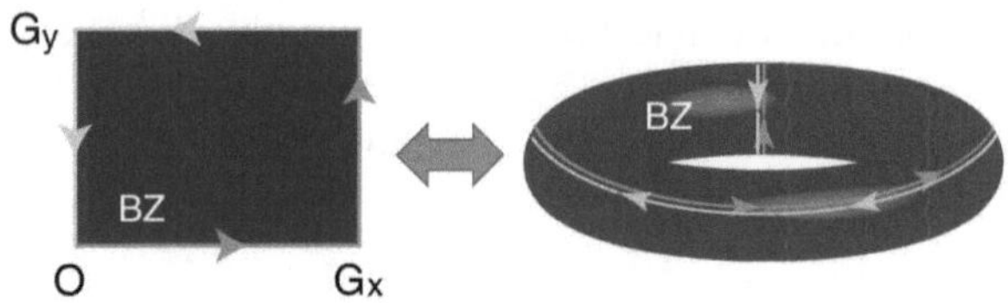

Fig. 2.1 Schematic the 2D BZ. $G_{x,y}$ denote the reciprocal wavevectors for the x and y directions. The *arrows* show the direction of the path integral for the Hall conductivity. In the calculation, the paths with opposite *arrows* cancel with each other, when the phase of the wavefunction is defined smoothly in the whole BZ

the wavefunction in the whole BZ. There is a simple way to determine the phase; we first fix a certain position $\mathbf{r} = (x, y)$, and we require a component of the state vector $|u_{\mathbf{k}}\rangle$ at the position $\mathbf{r}$, $u_{\mathbf{k}}(\mathbf{r}) = \langle \mathbf{r}|u_{\mathbf{k}}\rangle$, to be real. If $u_{\mathbf{k}}(\mathbf{r})$ has a non vanishing phase $g_{\mathbf{k}}$ which is assumed to be real and smooth, we perform the gauge transformation,

$$|u'_{\mathbf{k}}\rangle = \mathrm{e}^{-ig_{\mathbf{k}}}|u_{\mathbf{k}}\rangle, \tag{2.14}$$

where $\mathrm{e}^{ig_{\mathbf{k}}} = \frac{u_{\mathbf{k}}(\mathbf{r})}{|u_{\mathbf{k}}(\mathbf{r})|}$, and thus, we can make $u'_{\mathbf{k}}(\mathbf{r})$ to be real.

On the other hand, we cannot do this when the wavefunction vanishes for some $\mathbf{k}$. For simplicity, we assume that $u_{\mathbf{k}}(\mathbf{r})$ vanishes only at $\mathbf{k} = \mathbf{k}_0$. Then, the phase of $u_{\mathbf{k}}(\mathbf{r})$ cannot be determined at $\mathbf{k} = \mathbf{k}_0$; by the singularity, the phase of the wavefunction $u_{\mathbf{k}}(\mathbf{r})$ is determined except for $\mathbf{k} = \mathbf{k}_0$. Here, we again fix the phase of $u_{\mathbf{k}}(\mathbf{r})$ to be real as Eq. (2.14), for $\mathbf{k} \in \mathrm{BZ} - \{\mathbf{k}_0\}$. To make clear the argument, we divide the BZ into two, H_I and H_{II}; H_I contains a vicinity of $\mathbf{k} = \mathbf{k}_0$, and H_{II} is the complement of H_I in the BZ. Then, we choose another position $\mathbf{r}'$ such that the component of $u_{\mathbf{k}}(\mathbf{r}')$ does not vanish in the region H_I including $\mathbf{k} = \mathbf{k}_0$. Since $u_{\mathbf{k}}(\mathbf{r}')$ does not have a singularity in H_I, we can impose it to be real by using a real function $h_{\mathbf{k}}$ as

$$|u''_{\mathbf{k}}(\mathbf{r}')\rangle = \mathrm{e}^{-ih_{\mathbf{k}}}|u_{\mathbf{k}}(\mathbf{r}')\rangle, \tag{2.15}$$

where $\mathrm{e}^{ih_{\mathbf{k}}} = \frac{u_{\mathbf{k}}(\mathbf{r}')}{|u_{\mathbf{k}}(\mathbf{r}')|}$. Thus, we have a wavefunction whose phase is defined in the whole BZ in Fig. 2.2.

$$\begin{cases} |u_{\mathbf{k}\in H_I}\rangle = \mathrm{e}^{-ih_{\mathbf{k}}}|u_{\mathbf{k}}\rangle \\ |u_{\mathbf{k}\in H_{II}}\rangle = \mathrm{e}^{-ig_{\mathbf{k}}}|u_{\mathbf{k}}\rangle \end{cases}. \tag{2.16}$$

The remaining problem is the phase on the boundary between H_I and H_{II}. If the phases can be connected smoothly there, it is inconsistent with the initial assumption that we cannot define the phase to be smooth in the whole BZ. Therefore, there is a phase mismatch at the boundary ∂H,

$$|u_{\mathbf{k}\in H_I}\rangle = \mathrm{e}^{if_{\mathbf{k}}}|u_{\mathbf{k}\in H_{II}}\rangle, \tag{2.17}$$

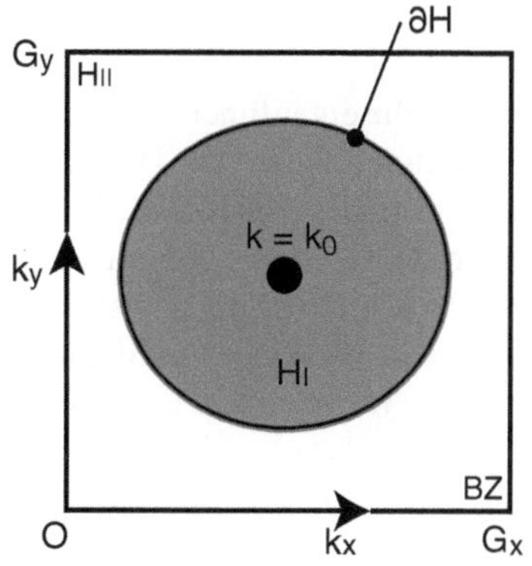

Fig. 2.2 Schematic of diagram of a phase of a wavefunction. To determine the phase on the entire BZ, it is divided into two regions. At $\mathbf{k} = \mathbf{k}_0$, there is a singularity of the phase for $u_{\mathbf{k}}(\mathbf{r})$

where $t_{\mathbf{k}} = g_{\mathbf{k}} - h_{\mathbf{k}}$. Because the connection depends on the gauge, we define the connections as $\mathbf{a}^I$ and $\mathbf{a}^{II}$ on H_I and H_{II}, respectively. The Hall conductivity coming from the two regions is expressed as

$$\begin{aligned}\sigma_{xy} &= \frac{e^2}{\hbar}\int_{\mathrm{BZ}} \mathrm{d}\mathbf{k}[\nabla_{\mathbf{k}} \times \mathbf{a}]_z \\ &= \frac{e^2}{\hbar}\int_{H_I} \mathrm{d}\mathbf{k}[\nabla_{\mathbf{k}} \times \mathbf{a}^I]_z + \frac{e^2}{\hbar}\int_{H_{II}} \mathrm{d}\mathbf{k}[\nabla_{\mathbf{k}} \times \mathbf{a}^{II}]_z \\ &= \frac{e^2}{h}\nu \end{aligned} \tag{2.18}$$

$$\nu = \frac{1}{2\pi}\int_{\partial H} \mathrm{d}\mathbf{k}\cdot(\mathbf{a}^I - \mathbf{a}^{II}) = \frac{1}{2\pi}\int_{\partial H} \mathrm{d}\mathbf{k}\cdot\nabla t_{\mathbf{k}}. \tag{2.19}$$

The constant ν defined in Eq. (2.19) is called the Chern number and must be integer because $t_{\mathbf{k}}$ comes from the gauge transformation, and its change around ∂H is an integer multiple of 2π. ν is a topological invariant; it does not change by any smooth deformation of the path ∂H. In addition, since the nontrivial winding of the phase difference $t_{\mathbf{k}}$ is supported by the existence of the singularity, it never changes from the integer by small external perturbation which does not close the band gap.

2.1.2 Adiabatic Charge Polarization by the Berry Phase

In the previous section, we have discussed the Hall current originating from the Berry phase. The Hall current emerges when the quantum states change with time; there is a correspondence between the perturbation and time evolution of the wavenumber. Here, we consider the charge polarization in the form of the Berry phase by an adiabatic change of external perturbations. The Schrödinger equation is given as

$$H(\mathbf{k},\lambda)|\psi_{\mathbf{k},\lambda}\rangle = E(\mathbf{k},\lambda)|\psi_{\mathbf{k},\lambda}\rangle, \tag{2.20}$$

where the eigenfunction $|\psi_{\mathbf{k},\lambda}\rangle$ has the Bloch form with $|\psi_{\mathbf{k},\lambda}\rangle = \mathrm{e}^{i\mathbf{k}\cdot\mathbf{r}}|u_{\mathbf{k},\lambda}\rangle$, and the wavefunction and the Hamiltonian explicitly depend on a parameter λ representing an external perturbation. λ changes from the initial state with $\lambda = \lambda_i$ to the final state with $\lambda = \lambda_f$ continuously. For example, when we apply pressure the system, the lattice structure is distorted weakly, and this affects the quantum states by the perturbation. Since the Bloch wavefunction spreads out over the system periodically, we use Wannier functions defined as

$$|\mathbf{R},\lambda\rangle = \frac{1}{\sqrt{N_0}}\sum_{\mathbf{k}} \mathrm{e}^{i\mathbf{k}\cdot(\mathbf{r}-\mathbf{R})}|u_{\mathbf{k},\lambda}\rangle, \tag{2.21}$$

where $\mathbf{R}$ specifies one of the lattice sites, and N_0 is the number of lattice sites. Then the polarization $\mathbf{P}$ in units of the electronic charge e is given as

$$\begin{aligned}\mathbf{P}_\lambda &= \langle\mathbf{R},\lambda|\mathbf{r}|\mathbf{R},\lambda\rangle \\ &= \sum_{\mathbf{k}} \mathbf{a}_{\mathbf{k},\lambda},\end{aligned} \tag{2.22}$$

where $\mathbf{a}_{\mathbf{k},\lambda}$ is the previously defined connection, and we used the relation

$$\begin{aligned}(\mathbf{r}-\mathbf{R})|\mathbf{R},\lambda\rangle &= \frac{1}{\sqrt{N_0}}\sum_{\mathbf{k}} -i(\nabla_{\mathbf{k}}\mathrm{e}^{i\mathbf{k}\cdot(\mathbf{r}-\mathbf{R})})|u_{\mathbf{k},\lambda}\rangle, \\ &= \frac{1}{\sqrt{N_0}}\sum_{\mathbf{k}}\Big[-i\nabla_{\mathbf{k}}(\mathrm{e}^{i\mathbf{k}\cdot(\mathbf{r}-\mathbf{R})}|u_{\mathbf{k},\lambda}\rangle) + i\mathrm{e}^{i\mathbf{k}\cdot(\mathbf{r}-\mathbf{R})}|\nabla_{\mathbf{k}}u_{\mathbf{k},\lambda}\rangle\Big].\end{aligned} \tag{2.23}$$

Therefore the polarization vector after the adiabatic change is given as

$$\Delta\mathbf{P} = \mathbf{P}_{\lambda_f} - \mathbf{P}_{\lambda_i} = \sum_{\mathbf{k}}(\mathbf{a}_{\mathbf{k},\lambda_f} - \mathbf{a}_{\mathbf{k},\lambda_i}), \tag{2.24}$$

Here, we assume that the wavefunctions for wavevectors different by a reciprocal lattice vector are equal, including the phase. Then the x-component of the polarization is given as

$$\begin{aligned}\Delta P^x &= -\frac{1}{2\pi}\int \mathrm{d}k_x(\tilde{a}^x_{k_x,\lambda_i} - \tilde{a}^x_{k_x,\lambda_f}) - \frac{1}{2\pi}\int \mathrm{d}\lambda(\tilde{a}^\lambda_{n\pi,\lambda} - \tilde{a}^\lambda_{n-\pi,\lambda}) \\ &= -\frac{1}{2\pi}\oint \mathrm{d}(k_x,\lambda)\cdot(\tilde{a}^x_{k_x,\lambda},\tilde{a}^\lambda_{k_x,\lambda}),\end{aligned} \tag{2.25}$$

where $\tilde{a}^{x,(\lambda)}_{k_x,\lambda_i} = \sum_{k_y,k_z} a^{x(\lambda)}_{\mathbf{k},\lambda_i}, a^\lambda_{\mathbf{k},\lambda_i} = \langle u_{\mathbf{k}\lambda}|i\partial_\lambda u_{\mathbf{k}\lambda}\rangle$, and the added term is zero by our assumption. Therefore the charge polarization is described by the contour integral of

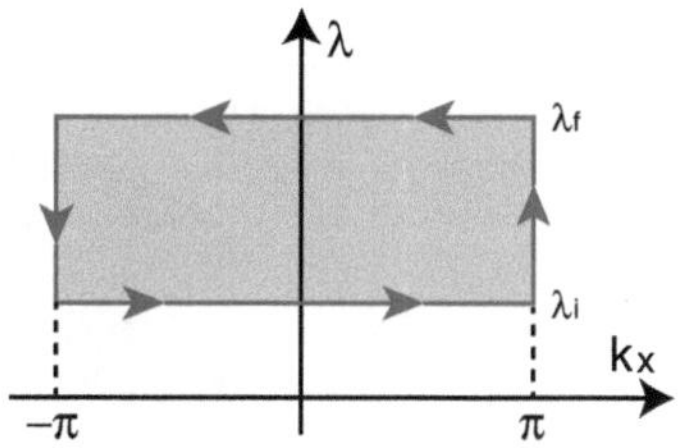

Fig. 2.3 Schematic of the path for the integral of the Berry connection

the Berry connection (Fig. 2.3). In addition, by the Stokes theorem, the polarization is represented by the Berry curvature as

$$\Delta P^x = -\frac{1}{2\pi}\oint \mathrm{d}k_x \mathrm{d}\lambda \tilde{b}^{x\lambda}_{k_x,\lambda_i}, \tag{2.26}$$

$$\tilde{b}^{x\lambda}_{k_x,\lambda_i} = \partial_x \tilde{a}^{\lambda}_{k_x,\lambda_i} - \partial_\lambda \tilde{a}^{x}_{k_x,\lambda_i}. \tag{2.27}$$

This polarization can be interpreted as the displacement of the center of the Wannier function, and the piezoelectric effect is studied based on this notion in detail [8, 9]. In particular, when the external perturbation is periodic, ΔP^x becomes an integer.

2.1.3 Laughlin's Gedanen Experiment

Laughlin gave an explanation of the quantization of the Hall conductivity [3], in terms of the gauge invariance. We consider a 2DEG on the xy plane subjected to a strong magnetic field along the z axis. The Hamiltonian is

$$H = \frac{1}{2m}(\mathbf{p} + e\mathbf{A})^2. \tag{2.28}$$

By using the Landau gauge,

$$\begin{cases} A_x = -By \\ A_y = 0 \end{cases}. \tag{2.29}$$

The system has translational symmetry along the x direction, and we impose the periodic boundary condition for the x axis, and the open boundary condition for the y axis with the size $L_x \times L_y$. Hence this system is regarded as a cylinder along the y axis. The Hamiltonian for the 2DEG is expressed as

$$\begin{aligned} H_{k_q} &= \mathrm{e}^{-ik_q x} H \mathrm{e}^{ik_q x} \\ &= \frac{1}{2m} p_y^2 + \frac{1}{2} m\omega_c^2 (y - k_q \ell^2)^2, \end{aligned} \tag{2.30}$$

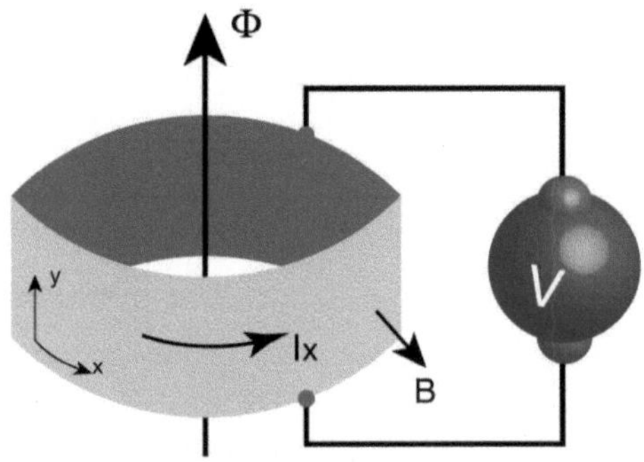

Fig. 2.4 Schematic of the geometry of the 2DEG system. The magnetic field is applied perpendicular to the cylinder. The system is periodic along the x axis, and the flux penetrates the cylinder. Along the y axis, the open boundary condition is imposed and the electronic field is applied. The Hall current is generated by the flux

where $\omega_c = \frac{eB}{m}$ is the cyclotron frequency, $\ell = \sqrt{\frac{\hbar}{eB}}$ is the magnetic length, and the wavenumber is given as $k_q = \frac{2\pi}{L_x} q$ with an integer q. From this Hamiltonian, the eigenstates has the same form as the harmonic oscillator, which has its center at $y_q = k_q \ell^2$. The eigenvalue forms the Landau level, given as

$$\varepsilon_n = \hbar\omega_c \left(\frac{1}{2} + n\right), \tag{2.31}$$

$$n = 0, 1, 2, \ldots, \tag{2.32}$$

which does not depend on the wavenumber. When the wavenumber is changed as $q \to q + 1$, while the energy spectrum is invariant, the center of the eigenstate shifts by $\delta y = \frac{2\pi\hbar}{eBL_x}$. Therefore, the system is invariant under the gauge transformation $\mathbf{A} \to \mathbf{A} + \delta\mathbf{A}$, where $\delta\mathbf{A} = (\frac{\delta\Phi}{L_x}, 0)$, and $\delta\Phi = \frac{2\pi\hbar}{e}$.

Now, we consider the Fermi energy is on the gap between the Landau levels ε_ν and $\varepsilon_{\nu+1}$. Due to the periodicity along the x axis, the Aharanov-Bohm flux along the cylinder axis is introduced (Fig. 2.4). The change of the energy by the flux is given as

$$\begin{aligned} \delta\langle H\rangle = \langle H(\mathbf{A} + \delta\mathbf{A})\rangle - \langle H(\mathbf{A})\rangle = -\frac{\delta\Phi}{L_x}\int_0^{L_x}\int_0^{L_y} d\mathbf{r}\, j_x \\ = -\delta\Phi I_x \end{aligned} \tag{2.33}$$

where $\langle \mathscr{O} \rangle$ denotes the average value of the operator $\mathscr{O}$, I_x is the current along the x axis, the current density is given as

$$j_x = -e\left\langle \frac{p_x + eA_x}{m} \right\rangle = -\left\langle \frac{\partial H}{\partial A_x} \right\rangle, \tag{2.34}$$

and we used the invariance of the sates under the flux change $\delta\mathbf{A}$. Therefore, the Hall current is given as the change of energy by the fictitious flux as

$$I_x = -\frac{\delta \langle H \rangle}{\delta \Phi}. \tag{2.35}$$

The change of the energy is estimated by the movement of the electron gas. Since the voltage $V(y)$ is applied along the y axis, the electrons are subjected to the electronic field $\frac{V(L_y)-V(0)}{L_y}$, and all of the electrons move along the y axis by $\delta y = \frac{2\pi\ell^2}{L_x}$ by the flux. Therefore $\delta\langle H\rangle$ is given as

$$\begin{aligned}\delta\langle H\rangle &= -e\delta y \times \frac{V(L_y) - V(0)}{L_y} \times \nu N_0 \\ &= -e\nu(V(L_y) - V(0))\end{aligned} \tag{2.36}$$

where $N_0 = \frac{L_x L_y}{2\pi\ell^2}$ is the degeneracy of one Landau level.[3] Therefore, we have the current as

$$I_x = \nu\frac{e^2}{h}(V(L_y) - V(0)), \tag{2.37}$$

with the quantized conductivity $\nu\frac{e^2}{h}$. The calculation of the change of the energy corresponds to the integral of the current density j_x with respect to the wavenumber k_x, since the momentum has the covariant form, $\mathbf{k} + e\mathbf{A}$. In another perspective, charges move along the y axis by δy by the flux insertion, and therefore the charge polarization occurs.

In this context, we discuss the charge polarization previously mentioned in Sect. 2.1.2. We consider the 2DEG system with the periodic boundary condition along the x and y axes. The system can be regarded as a torus, and the flux penetrates along the y axis. By the flux $\delta\Phi$, the variation of the gauge is given as $\delta\mathbf{A} = (0, \frac{\delta\Phi}{L_y})$, and a change of the charge polarization at k_y along the x axis becomes

$$\Delta P^x(k_y) = \frac{1}{2\pi}\int \mathrm{d}k_x (a_{x\mathbf{k}+e\delta\mathbf{A}} - a_{x\mathbf{k}}). \tag{2.38}$$

Since the flux causes the momentum shift, the total polarization is given as

$$\begin{aligned}\Delta P^x &= \sum_{k_y=0}^{G_y} \Delta P^x(k_y) = \frac{1}{2\pi}\int \mathrm{d}k_x (a_{x\mathbf{k}+(0,G_y)} - a_{x\mathbf{k}}) \\ &= -\frac{1}{2\pi}\int \mathrm{d}\mathbf{k} b_z(\mathbf{k}).\end{aligned} \tag{2.39}$$

[3] When the Landau level is filled, the electron density per Landau level is given as $\frac{N_0}{L_x L_y} = \frac{eB}{h} = \frac{1}{2\pi\ell^2}$.

Therefore, the change of the charge polarization by flux insertion is also described by the Chern number.

We have seen the states move in momentum space by the flux insertion. This movement and the electronic current imply existence of the electronic states crossing the Fermi energy, which cannot be seen in the bulk states. Indeed, these hidden states are localized at edge, and this fact is consistent with the quantized conductivity that does not depend on the size of the system. Thus, we conclude that the Chern number is an indication of the boundary states carrying the quantized conductivity. While the Chern number is defined as a quantity in the bulk with translational symmetry, remarkably, we can extract information of the boundary states from it. This is a basic notion of the bulk-boundary correspondence, which has been applied to broad classes of topological insulators.

2.1.4 Physical Picture of the Edge State

The topological order in the QH effect manifests itself in the existence of the edge states, which is localized on the edge of system, and the quantized conductivity is carried by the edge channel. In this section we explain an intuitive picture of edge states proposed by Halperin and Büttiker [10, 11]. We assume a 2DEG system in the xy plane with a magnetic field $\mathbf{B}$ along the z axis, and an electric field $\mathbf{E}$ is applied along the y axis (Fig. 2.5a). The classical velocity of the center of the cyclotron motion $\mathbf{v}_c$ is given as

$$\mathbf{v_c} = \frac{\mathbf{E} \times \mathbf{B}}{B^2}. \tag{2.40}$$

For the n-th Landau levels, the energy of the edge states E_n is described as

$$E_n = -eW(\mathbf{r}) + \hbar\omega_c\left(\frac{1}{2} + n\right), \tag{2.41}$$

where ω_c is the cyclotron frequency, and $-eW(\mathbf{r})$ is the confinement potential at the position $\mathbf{r}$ (Fig. 2.5b). The density of states in each Landau level per unit area is $\frac{Be}{h}$. Therefore, the Hall current density j_x^n for the n-th Landau level is given as

$$j_x^n = -e\left(\frac{Be}{h}\right)\theta\left(\mu - eW(\mathbf{r}) + \hbar\omega_c\left(\frac{1}{2} + n\right)\right)\left(-\frac{1}{B}\frac{\partial W}{\partial y}\right), \tag{2.42}$$

where we use $v_{cx} = \frac{E_y}{B} = -\frac{1}{B}\frac{\partial W}{\partial y}$ due to the confinement, and μ is the chemical potential. The Hall current I_x^n at the edge is given as

$$I_x^n = \frac{e^2}{h}\int_0^{L_y} \mathrm{d}y\theta\left(\mu - eW(\mathbf{r}) + \hbar\omega_c\left(\frac{1}{2} + n\right)\right)\left(\frac{\partial W}{\partial y}\right). \tag{2.43}$$

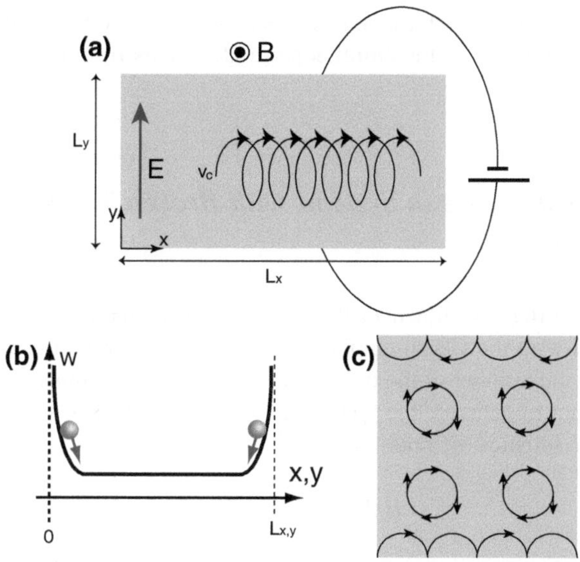

Fig. 2.5 **a** Schematic of a 2DEG system. The magnetic field B is applied along the z axis, and the electric field E is along the y axis. The *curve* illustrates a cyclotron motion with the classical velocity v_c. **b** Schematic of the confinement potential W. W is divergent outside of the system. The *balls* represent electrons confined within the system, and the *arrows* show the directions of the force due to the effective electric field $\frac{\partial w}{\partial \mathbf{r}}$. **c** Schematic of the cyclotron motion in a 2DEG system. The cyclotron motion is illustrated as *circles* whose *arrows* represent the direction of the motion. The neighboring motions cancel each other, and only the skipping orbits at the edge remain

When $\mu > E_n(\mathbf{r})$ for any $\mathbf{r} \in [0, L_x] \times [0, L_y]$, the Hall current is given as

$$I_x^n = \frac{e^2}{h}(W(L_y, x) - W(0, x)) = \frac{e^2}{h}V, \tag{2.44}$$

where we use the fact that the difference of the confinement potential corresponds to the electric voltage V in the y direction. When the n-th Landau level with $0 \leq n \leq \nu - 1$ is filled, the Hall current I_x is given by

$$I_x = \nu\frac{e^2}{h}(W(L_y, x) - W(0, x)) = \nu\frac{e^2}{h}V, \tag{2.45}$$

which is a well-known result for the integer QH effect. In this case, each Landau level is shown to have the Chern number unity. Then we can see that Eq. (2.45) is a special case of Eq. (2.7).

We note that the edge mode is interpreted also from the picture of a collection of the cyclotron motions. In the collection of the cyclotron motions under the magnetic field, they cancel each other in the bulk, but do not on the edge. Only the cyclotron motions forming skipping orbits along the edges remain on the edges, and they make the edge current called the chiral edge states (Fig. 2.5c). From this intuitive picture, we see that the direction of the chiral edge mode results from that of the cyclotron motion.

2.1.5 Berry Curvature in Systems with Broken Time-Reversal Symmetry

So far, the QH effect is shown in the 2DEG in a magnetic fields. Each Landau level is labeled with the Chern number $n_c = 1$, and Chern numbers of the system corresponds to the number of the occupied Landau levels. In this case, the magnetic field is important to generate the QH effect. In a time-reversal invariant system the Hamiltonian H satisfies the relation

$$\Theta H(\mathbf{k})\Theta^{-1} = H(-\mathbf{k}), \tag{2.46}$$

where $\mathbf{k}$ is the wavevector, and the time-reversal operator $\Theta = i\sigma_y K$ is represented by the product of the y-component of the Pauli matrix σ_y acting on spin degrees of freedom and the complex conjugation operator K. If the system has the time-reversal symmetry, the Chern number ν vanishes because the Berry curvature satisfies the relation (Sect. 6.1.1),

$$\mathbf{b}(\mathbf{k}) = -\mathbf{b}(-\mathbf{k}). \tag{2.47}$$

Conversely, a nonzero Chern number requires breaking of time-reversal symmetry. In the normal QH effect, magnetic field s are used to break time-reversal symmetry. However magnetic fields are not necessary to realize the QH effect, as in the case for the Haldane model [12]. On the other hand, in time-reversal-invariant systems, we need another scheme for topological characterization in terms of a new topological invariant. It leads to the Z_2 topological number, as is introduced in the next section.

2.2 Topological Insulator

Topological insulators are new states of matter, which are proposed by Kane and Mele, and by Bernevig and Zhang [13–18]. In TIs, the bulk is gapped while the boundary of the system is metallic and they are realized in the presence of time-reversal symmetry. Strong spin-orbit interaction (SOI) is required to realize TIs.

The two-dimensional (2D) TI was first proposed in graphene, and the gap is formed by the spin-orbit interaction. However, the SOI is too small to realize the 2DTI in graphene in experiments. After that proposal, Bernevig et al. theoretically predicted realization of the 2DTI in HgTe quantum well in 2006 [15], and in 2007 König

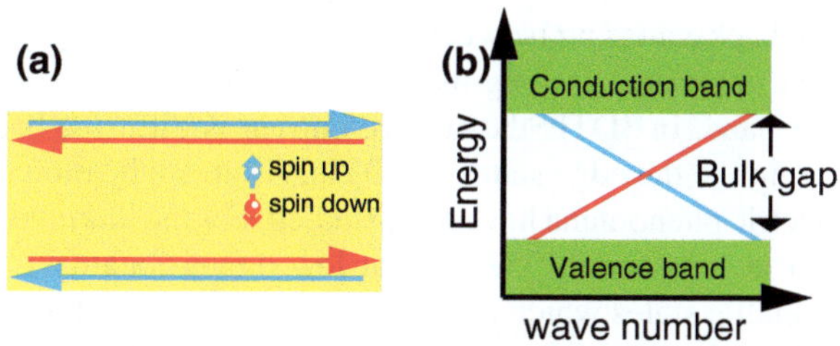

Fig. 2.6 Schematics illustrating edge states of 2DTIs. **a** The edge current of the 2DTI in the ribbon geometry, **b** The dispersion of the edge states is shown as *blue* and *red lines*. There are two states with opposite directions of velocities and opposite spins, forming a Kramers pair

et al. observed the 2DTI through the quantized edge conductivity in the quantum well [19]. In 2DTIs, there are edge states consisting of a pair of spin-filtered states, whose spins and velocities are opposite to each other due to the Kramers's theorem (Fig. 2.6a). Such edge states consisting of counterpropagating states are called helical edge states. The 2DTI can be understood as a superposition of two QH systems with opposite magnetic fields for the two spin states (Fig. 2.6b).

In 2007 three-dimensional (3D) TIs were predicted theoretically [16, 17, 20] and $Bi_{1-x}Sb_x$ was predicted as a 3DTI [17]. Subsequently they were realized in experiments: $Bi_{1-x}Sb_x$ [21], Bi_2Se_3 [22, 23] and Bi_2Te_3 [24]. Since then, the search for 3DTI has been ongoing actively. An effective theory for the surface states of the 3DTI is described by the massless Dirac equation. The dispersion is linear in the wavevector, and typically the spin is perpendicular to the wavevector. There are two possibilities for the relation between the spin and wavevector (Fig. 2.7), called chirality. Materials which were identified as 3DTIs such as Bi_2Se_3, and Bi_2Te_3 have the chirality illustrated in Fig. 2.7a. A recent experiment shows that the surface states in $Bi_4Se_{2.6}S_{0.4}$ have the opposite chirality from these materials (Fig. 2.7b) [25].

Transport phenomena by these boundary states in TI are unique. In the edge states of 2DTIs, the back-scattering is prohibited, and therefore the edge channel perfectly conducts. This is because the two states within the edge channels form a

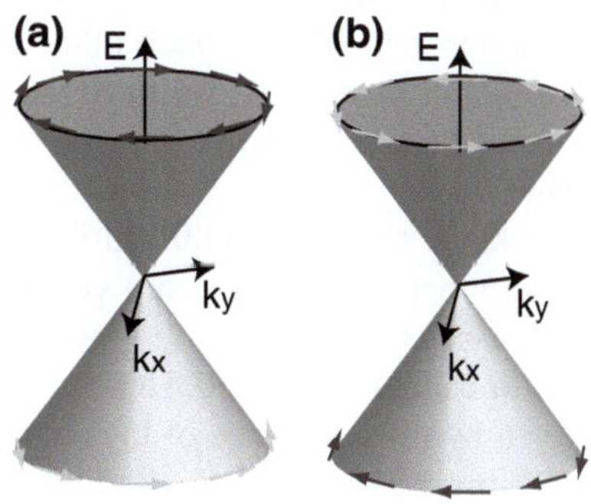

Fig. 2.7 Schematics of the dispersion of the surface states (**a**, **b**). The *circle* shows the Fermi surface of the surface states above the Dirac point, and the *arrows* pointing along the Fermi surface are spins. **a** and **b** correspond to opposite chiralities

Kramers pair, and backscattering between them is prohibited; the gapless edge states are robust against impurities as long as time-reversal symmetry is preserved [26, 27]. On the other hand, in 3DTIs the surface current is diffusive because impurity scatterings occur unlike the edge states in 2DTIs. As an application of the transport in TIs, thermoelectric phenomena have been studied. For the thermoelectronic figure of merit [28, 29], it is generally known that low-dimensionality is favorable [32, 33]. There are metallic states which are localized at the edges of TIs, and in several papers, enhanced thermoelectric transport has been reported by using the edge states of TIs [30, 31, 34].

2.2.1 Topological Invariant Under Time-Reversal Symmetry

Systems under time-reversal symmetry are characterized by the Z_2 topological numbers. The Z_2 topological number was first introduced in 2D systems [13–16, 35]. Then later, the theory was extended to 3D [17]. In the following, we explain the idea of the Z_2 topological number.

By the time-reversal symmetry, the Bloch wavefunctions $|u(\mathbf{k})\rangle$ and $|\Theta u(\mathbf{k})\rangle$ are degenerate, and orthogonal to each other (Sect. 6.2). This is the Kramers theorem; the dispersion is symmetric with respect to $\mathbf{k}$ because the wavevector of the states $|\Theta u(\mathbf{k})\rangle$ is $-\mathbf{k}$, and therefore the states at $\mathbf{k}$ and $-\mathbf{k}$ have opposite spins from each other. The energy spectra necessarily cross and the states are doubly degenerate by Kramers's theorem at time-reversal invariant momenta (TRIM) $\mathbf{k} = \boldsymbol{\Gamma}$ where $\boldsymbol{\Gamma} = -\boldsymbol{\Gamma}\ (\mathrm{mod}\,\mathbf{G})$ and $\mathbf{G}$ is a reciprocal lattice vector; in other words, the degeneracy at the TRIM is protected as long as time-reversal symmetry is preserved.

In systems with time-reversal symmetry, the Chern number vanishes because Berry curvatures contributed from $|u\rangle$ and $|\Theta u\rangle$ cancel each other. We first consider a special situation where the states are eigenstates of the spin along z, $s_z = \pm$ and the two subspaces with $s_z = +$ and $s_z = -$ are decoupled. Therefore $|u\rangle$ and $|\Theta u\rangle$ have opposite spins to each other, and are labeled with $s_z = \pm$. Then edge currents from $s_z = \pm$ states run along opposite directions with opposite spins due to the SOI as the driving force [36–41], which does not break time-reversal symmetry. Therefore in this simple case we can construct a topological invariant (Chern number) for each subspace with $s_z = \pm$. Namely the two subspaces from $s_z = \pm$ have Chern numbers with opposite signs to each other. When these Chern numbers are not zero, as we learned in Sect. 2.1.4, we have the charge polarizations when the wavefunction is changed adiabatically by a reciprocal lattice vectors, and it leads to the topological edge states.

This notion is generalized in terms of time-reversal polarization [35] when s_z is no longer conserved; in a general system with time-reversal symmetry, the two subspaces $s_z = +$ and $s_z = -$ are not decoupled, and the Chern numbers within each sector cannot be defined. Hence, in general cases, topological invariants are directly given by the difference of the charge polarizations between the pairs connected by time-reversal operation: $|u\rangle$ and $|\Theta u\rangle$. For 2D systems, by calculating the difference of the charge polarizations by using Eq. (2.39), under a change of wavevector by a half of the reciprocal lattice vector along the x axis, the Z_2 topological number ν is defined as

$$\nu \equiv \frac{1}{2\pi}\left[\int_{\mathbf{BZ}_{1/2}} \mathrm{Tr}(\mathscr{F}) - \oint_{\partial\mathbf{BZ}_{1/2}} \mathrm{Tr}(a)\right] \pmod 2 \tag{2.48}$$

where $\mathrm{BZ}_{1/2}$ means a half of the BZ. The topological number takes $\nu = 1$ for topologically nontrivial systems or $\nu = 0$ for ordinary insulators. This expression coincides with the above discussion because the integration of $\mathrm{Tr}(\mathscr{F})$ over half of the BZ means the polarization contributed from one of the Kramers pair. We note that the definition of the Z_2 topological number Eq. (2.48) can be extended to 3DTIs.

The Z_2 topological number can be represented by a more convenient form. To this end, we define the following $2N \times 2N$ matrix as

$$w_{mn}(\mathbf{k}) = \langle u_{-\mathbf{k},m}|\Theta|u_{\mathbf{k},n}\rangle, \tag{2.49}$$

where m, n are band indices, which run over occupied bands, and $2N$ is the total number of occupied bands. At each TRIM $\Gamma_{i=1,2,\ldots,M}$, an index $\delta_i = \pm 1$ is defined as

$$\delta_i \equiv \frac{\sqrt{\det\left[w(\boldsymbol{\Gamma}_i)\right]}}{\mathrm{Pf}\left[w(\boldsymbol{\Gamma}_i)\right]}, \tag{2.50}$$

and then the Z_2 topological number ν is defined as

$$(-1)^{\nu} \equiv \prod_{i\in\mathrm{TRIM}}^{M} \delta_i, \tag{2.51}$$

where M is the number of TRIM in the BZ e.g., $M = 4$ in 2D, $M = 8$ in 3D.[4] Due to $\mathrm{Pf}[w(\boldsymbol{\Gamma}_i)]^2 = \det[w(\boldsymbol{\Gamma}_i)]$ there seems to be an ambiguity in the sign of $\sqrt{\det[w(\boldsymbol{\Gamma}_i)]}$ relative to $\mathrm{Pf}[w(\boldsymbol{\Gamma}_i)]$. It is uniquely determined by defining the wave-function smoothly between TRIM in the BZ. For this purpose we define the function as

$$e^{i\theta(\mathbf{k})} \equiv \frac{\det\left[w(\mathbf{k})\right]}{\left|\det\left[w(\mathbf{k})\right]\right|}. \tag{2.52}$$

We consider a path between two TRIM $\boldsymbol{\Gamma}_{a,b}$ in the BZ, and vary the phase $e^{i\theta(\mathbf{k})}$ smoothly along the path. At $\boldsymbol{\Gamma}_a$ and $\boldsymbol{\Gamma}_b$, we define $\theta_0(\boldsymbol{\Gamma}_a)$ and $\theta_0(\boldsymbol{\Gamma}_b)$ as

$$e^{i\frac{\theta_0(\boldsymbol{\Gamma}_i)}{2}} \equiv \frac{\mathrm{Pf}[w(\boldsymbol{\Gamma}_i)]}{|\mathrm{Pf}[w(\boldsymbol{\Gamma}_i)]|}. \tag{2.53}$$

---

[4] In 1D the Z_2 topological number cannot be defined. For example, by a gauge transformation $|u\rangle \to e^{ik/(2\Gamma)}|u\rangle$ where Γ is a half of the reciprocal lattice vector, the Z_2 topological number changes as $(-1)^{\nu} \to -(-1)^{\nu}$.

Then, by going from $\boldsymbol{\Gamma}_a$ to $\boldsymbol{\Gamma}_b$, the gained phase of the determinant of $w(\mathbf{k})$ is

$$\theta_0(\boldsymbol{\Gamma}_b) - \theta_0(\boldsymbol{\Gamma}_a) + 2\pi \Delta_{ab}, \tag{2.54}$$

where Δ_{ab} is an integer. The adiabatic transition of the phase is expressed as

$$\sqrt{\frac{\det[w(\boldsymbol{\Gamma}_b)]}{|\det[w(\boldsymbol{\Gamma}_b)]|}}\sqrt{\frac{|\det[w(\boldsymbol{\Gamma}_a)]|}{\det[w(\boldsymbol{\Gamma}_a)]}} = \mathrm{e}^{i\frac{\theta_0(\boldsymbol{\Gamma}_b)-\theta_0(\boldsymbol{\Gamma}_a)}{2}+\pi i \Delta_{ab}} \tag{2.55}$$

From Δ_{ab}, we define another index related with the time-reversal polarization π_{ab} as

$$\pi_{ab} \equiv \delta_a \delta_b = \mathrm{e}^{\pi i \Delta_{ab}}. \tag{2.56}$$

When the phase turns even (odd) times in the Gauss plane $\delta_a\delta_b = +1(-1)$. By using Eq. (2.56), the Z_2 topological number (Eq. 2.51) is calculated. Figure 2.8 shows an example of the calculation of ν in a 2D system with a function $R(\mathbf{k}) = \mathrm{e}^{i\theta}$. In this case we have $\delta_1\delta_2 = -1$ and $\delta_3\delta_4 = +1$ for the paths $\Gamma_1 \to \Gamma_2$ and $\Gamma_3 \to \Gamma_4$, respectively; it leads to $\nu = 1$, and the system is a TI. There are edge states which cross the Fermi energy.

2.2.1.1 Z_2 Topological Number in 3D and Phase Transitions

In 3Ds, there are eight TRIM in the BZ which are described as

$$\Gamma_{\mathbf{n}} = \frac{1}{2}(n_1\mathbf{G}_1 + n_2\mathbf{G}_2 + n_3\mathbf{G}_3), \tag{2.57}$$

where $n_{i=1,2,3} = 0, 1$, The Z_2 topological numbers $(\nu_0; \nu_1, \nu_2, \nu_3)$ are defined as

$$(-1)^{\nu_k} \equiv \prod_{n_k=1; n_{j\neq k}} \delta_{\mathbf{n}}, \tag{2.58}$$

where $k = 1, 2, 3$, and ν_0 is described by Eq. (2.51) for $M = 8$. For $\nu_0 = 1$ the system is in the topologically non-trivial phase, called a strong TI, which has an odd number of Dirac cones as the surface states. Strong TIs are also simply called as TI, because the novel surface states are guaranteed by the Kramers theorem. On the other hand for $\nu_0 = 0$, there are an even number of Dirac cones as the surface states, and the system is called a weak TI. In weak TIs, the systems can be regarded as stacked layers of 2DTI. By using the indices (ν_1, ν_2, ν_3) the normal vector of the layers constituting the weak TI is described as

$$\mathbf{G}_\nu = \nu_1\mathbf{G}_1 + \nu_1\mathbf{G}_3 + \nu_3\mathbf{G}_3. \tag{2.59}$$

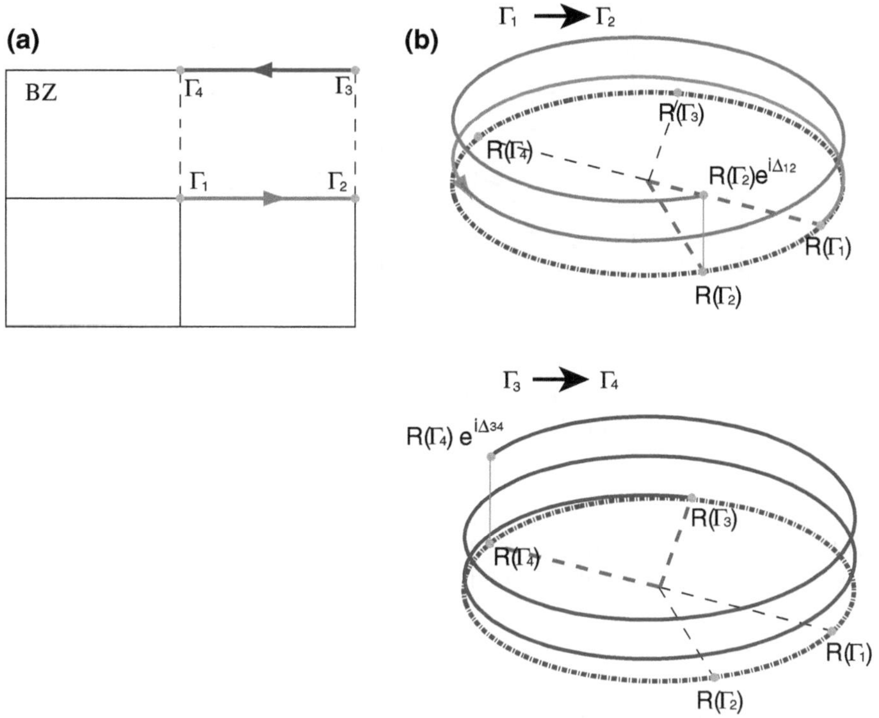

Fig. 2.8 Schematic example of the calculation of ν in a 2D system. $R(\mathbf{k})$ is the phase of Eq. (2.52), plotted as $e^{iR(\mathbf{k})}$ in the complex plane. **a** The BZ of a 2D system and TRIM. **b** The change of the phase (Eq. 2.52) in each path between TRIM. In this case we have $\delta_1\delta_2 = -1$ and $\delta_3\delta_4 = +1$ for the paths $\Gamma_1 \to \Gamma_2$ and $\Gamma_3 \to \Gamma_4$, respectively; namely $\nu = 1$, and the system is a strong TI

The weak TI is similar to ordinary insulators, because it includes trivial band insulators, and even if there are nonzero number of Dirac cones as the surface states, the gapless states are fragile against impurities or defects.

To illustrate the surface states of strong and weak TIs, we define an index associated with each surface TRIM as

$$\pi^z_{n_1,n_2} \equiv \delta_{(n_1,n_2,0)}\delta_{(n_1,n_2,1)}. \tag{2.60}$$

In the 2D, we have explained that bound states cross the Fermi surface between TRIM with different time-reversal polarizations. Then the Fermi surface on the (001) surface is illustrated as Fig. 2.9 schematically. In Fig. 2.9a the system is the strong TI with $(\pi^z_{0,0}, \pi^z_{1,0}, \pi^z_{01}, \pi^z_{11}) = (1, 1, 1, -1)$. In this case there is a single-Dirac cone whose Fermi surface encloses $\bar{\Gamma}_{11}$. On the other hand for $(\pi^z_{0,0}, \pi^z_{1,0}, \pi^z_{01}, \pi^z_{11}) = (-1, 1, 1, -1)$ (Fig. 2.9b), the system is the weak TI with an even number of Dirac cones.

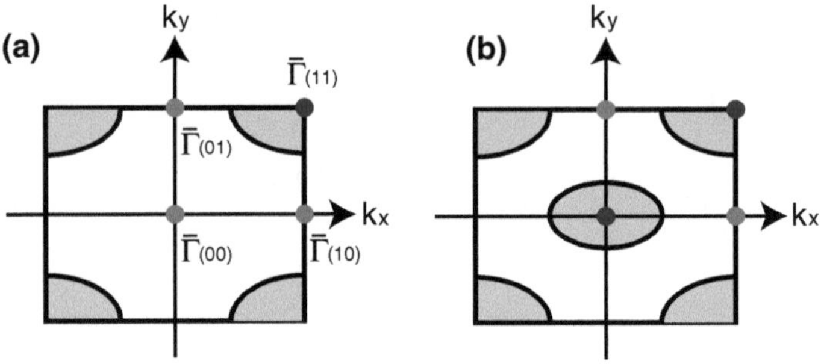

Fig. 2.9 Schematic examples of surface Fermi surfaces for the (001) surface. $\bar{\Gamma}_{n_1 n_2}$ are TRIM of the surface BZ. *Curves* represent possible Fermi surfaces in the strong TI (**a**) $(\pi^z_{0,0}, \pi^z_{1,0}, \pi^z_{01}, \pi^z_{11}) = (1, 1, 1, -1)$, and in the weak TI (**b**) $(\pi^z_{0,0}, \pi^z_{1,0}, \pi^z_{01}, \pi^z_{11}) = (-1, 1, 1, -1)$. *Blue* and *red dots* represent TRIM with $\pi^z = +1$ and $\pi^z = -1$ respectively

The Z_2 topological number ν is also discussed in the context of the topological field theory without interactions [42, 43]. In the theory, the effective action for electromagnetic fields S_{eff} in TIs is expressed as

$$S_{\text{eff}} = S_{\text{Maxwell}} + S_{\text{topo}} = \int \mathrm{d}^3x\mathrm{d}t \left[\frac{1}{16\pi} F^{\mu\nu} F_{\mu\nu} + \frac{\alpha\theta}{32\pi} \varepsilon^{\mu\nu\sigma\tau} F^{\mu\nu} F_{\sigma\tau} \right], \tag{2.61}$$

where $F^{\mu\nu}$ is the field strength of the electromagnetic fields, S_{Maxwell} describes the conventional electromagnetic action, S_{topo} is characteristic in TIs, $\varepsilon^{\mu\nu\sigma\tau}$ is the antisymmetric tensor, α is the fine-structure constant and $\theta = \pi\nu$. From S_{topo}, the Lagrangian $\mathscr{L}_{\text{topo}}$ is described as

$$\mathscr{L}_{topo} = \theta \frac{e^2}{2\pi h} \mathbf{E} \cdot \mathbf{B}. \tag{2.62}$$

This causes the magnetoelectric effect [42, 44], where the time-reversal symmetry needs to be broken. Due to the correspondence between ν and θ, trivial insulators have $\nu = 0$ and S_{topo} vanishes. In this theory ν is described as

$$\nu = \frac{\theta}{\pi} \equiv \frac{1}{4\pi^2} \int \text{Tr} \left(a \wedge \mathrm{d}a + i\frac{2}{3} a \wedge a \wedge a \right). \tag{2.63}$$

This is also known as the Chern-Simons term in the quantum field theory.

2.2.1.2 Z_2 Topological Number with Inversion Symmetry

When the system is invariant under the parity transformation (space inversion), the index δ_i is given as [45]

$$\delta_i = \prod_m^N \xi_{2m}(\boldsymbol{\Gamma}_i), \tag{2.64}$$

where $\xi_{2m}(\boldsymbol{\Gamma}_i)$ is the eigenvalue of the parity operator P of the $2m$-th band at a TRIM Γ_i and the takes the value $\xi_{2m}(\boldsymbol{\Gamma}_i) = \pm 1$. This formula (2.64) was proposed by Fu and Kane. From this formula, $Bi_{1-x}Sb_x$ for $0.07 < x < 0.22$ was predicted as a TI. At each TRIM, each band is doubly degenerate due to the Kramers theorem; the $2m$-th and $2m+1$-th bands have the same eigenvalue. This representation makes clear the condition when the topological phase transition occurs. We consider a case where there are $2N$ filled bands and the Fermi energy is in the gap between the $2N$-th and $2(N+1)$-th bands. A 2×2 Hamiltonian H of the $2N$-th and $2(N+1)$-th bands on a TRIM Γ_i are represented as

$$H = \begin{pmatrix} \Delta & 0 \\ 0 & -\Delta \end{pmatrix}, \tag{2.65}$$

where $\pm\Delta$ are the energy eigenvalues of the bands. Due to the degeneracy, we omit the eigenstates of $2N - 1$-th and $2N + 1$-th bands from the representation of the Hamiltonian. The eigenstates are

$$|\psi_0\rangle = \begin{pmatrix} 1 \\ 0 \end{pmatrix}, \quad |\psi_1\rangle = \begin{pmatrix} 0 \\ 1 \end{pmatrix}, \tag{2.66}$$

where $|\psi_a\rangle$ represents the eigenstate of the $2(N+a)$-th band for $a = 0, 1$, with the eigenvalue of the parity as P_a (Fig. 2.10). We treat Δ as a tunable parameter, and the Fermi energy is assumed to be zero.[5] δ_i becomes

$$\delta_i = \begin{cases} \prod_m^{N-1} \xi_{2m}(\boldsymbol{\Gamma}_i) P_1 & (\Delta > 0), \\ \prod_m^{N-1} \xi_{2m}(\boldsymbol{\Gamma}_i) P_0 & (\Delta < 0). \end{cases} \tag{2.67}$$

When P_0 and P_1 are of opposite signs, the time-reversal polarization changes with the sign of Δ. We further assume that the tuning of Δ does not affect any other states on TRIM. In this case when Δ gets across zero, a phase transition occurs. Namely, when the Fermi energy is sandwiched between bands with opposite eigenvalues of the parity,[6] either of the following two cases occurs; (i) the phase is a TI or (ii) the phase is a normal insulator and it becomes a TI if the gap is inverted. The Z_2 topological numbers were also calculated by this formula in Bi_2Te_3 and Bi_2Se_3 [46]. The calculation predicted that they are TIs with a single-Dirac cone on the Fermi surface, which well agrees with experiments in these materials [22–24].

---

[5] We assume the tuning affect only the four bands around the Fermi energy.

[6] For example ψ_1 consists of s-orbitals, and ψ_2 consists of p-orbitals.

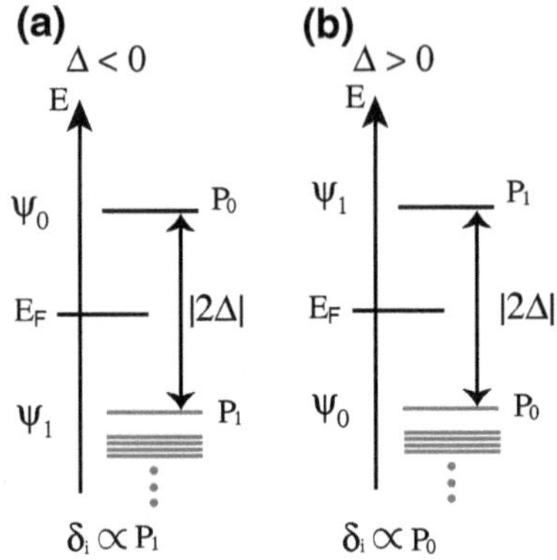

Fig. 2.10 Schematics of energy spectra. E_F is the Fermi energy, δ_i is the index related with the time-reversal polarization, $\psi_{0,1}$ and $P_{0,1}$ are states of the $2N$-th and $2(N+1)$-th bands and the parity eigenvalues. $\delta_i \propto P_1$ for **a** $\Delta > 0$, and $\delta_i \propto P_0$ for **b** $\Delta < 0$. When P_0 and P_1 have opposite signs, the index δ_i changes the sign by tuning Δ from $\Delta < 0$ to $\Delta > 0$

2.2.2 Surface Effective Hamiltonian

When a system is in the Z_2 non-trivial phase, the surface states can be obtained from the bulk Hamiltonian by imposing appropriate boundary conditions. This can also be done by using a low-energy effective model [46–48]. Near the TRIM where the surface states are degenerate, called the Dirac point, the surface Hamiltonian H_s and its eigenvalue E_s are expressed as

$$H_s(\mathbf{k}) = v(\sigma_x k_y - \sigma_y k_x), \tag{2.68}$$
$$E_s = s|v\mathbf{k}| \tag{2.69}$$

where v is the Dirac velocity and the sign of $s(=\pm)$ corresponds to the upper or lower Dirac cones.[7] The eigenstate ψ_s is given as

$$\psi_s = \frac{1}{\sqrt{2}}\begin{pmatrix} is \\ \frac{v}{|v|}e^{i\phi_\mathbf{k}} \end{pmatrix} e^{i\mathbf{k}\cdot\mathbf{r}}, \tag{2.70}$$

where $\phi_\mathbf{k} = \arctan\left(\frac{k_y}{k_x}\right)$ and $\mathbf{r}$ represent the coordinate on the surface. The sign of the velocity $\frac{v}{|v|}$ corresponds to the chirality, (a) or (b) in Fig. 2.7. They are topologically distinguished when the system has mirror symmetry. This correspondence will be discussed in Chap. 3 in detail.

[7] The Hamiltonian (Eq. 2.68) is transformed into $v(\boldsymbol{\sigma}\cdot\mathbf{k})$ by the unitary transformation, which is known as the massless Dirac Hamiltonian. Therefore we call Eq. (2.68) as the Dirac Hamiltonian.

2.3 Topological Flat-Band States in Honeycomb Lattice

Previously, we have seen emergence of edge states in topological insulators. Here, we review flat-band states on edges. Manifestation of the edge states in this class is supported by metallic bulk states. A similar type of the flat-band states will be discussed in Chap. 4, and the origin of the edge states is explained. Flat bands have been studied particularly in the context of possible ferromagnetism driven by interactions, as was proposed by Lieb [49], and successively by Mielke and Tasaki [50–53]. On the other hand, from the research on graphene [54] it is known that the tight-binding model with nearest-neighbor hopping on a honeycomb lattice with a zigzag edge exhibits flat-band edge states [55], and its origin is topologically interpreted [56]. In addition, the flat-band edge states can be seen in the 3D diamond lattice model [57].

Here we review the flat-band states in the honeycomb lattice model and discuss their topological origin. In the dispersion of a graphene ribbon with zigzag edges, the flat-band edge states appear between the wavenumbers corresponding to the projection of Dirac points at K and K′. In contrast, there are no flat-band edge states in the graphene ribbon with armchair edges, because in the projection of the dispersion, Dirac cones at the K and K′ points overlap each other.

When the hopping of the tight-binding model on the honeycomb lattice becomes anisotropic, the Dirac points in the bulk BZ move away from the K and the K′ points. Moreover, when the anisotropy is sufficiently large, the two Dirac points meet and the bulk dispersion relation becomes linear in one direction and quadratic in the other [58]. In that case, the flat-band edge states cover the whole one-dimensional (1D) BZ [59]. With a further increase of the anisotropy, the bulk becomes gapped while the completely flat band remains in the edge BZ.

2.3.1 Dispersion of the Honeycomb Lattice Model

We first review the flat-band edge states on the honeycomb-lattice structure shown in Fig. 2.11a, and study the completely flat band for the models with anisotropy, which has been studied by Delplace et al. [59]. We consider a tight-binding Hamiltonian on this lattice,

$$H = \sum_{\langle ij \rangle} c_i^\dagger t_{ij} c_j, \tag{2.71}$$

where t_{ij} is the hopping integral along the nearest-neighbor bond vector $\boldsymbol{\tau}_a$, and c_i ($c_i^\dagger$) is the annihilation (creation) operator of the electron. We treat the hopping integral t_{ij} as a real positive parameter, and it is labeled with the vectors $\boldsymbol{\tau}_a$ as t_a. The bulk Hamiltonian matrix $H_\mathrm{b}(\mathbf{k})$ at wavevector $\mathbf{k}$ is given as

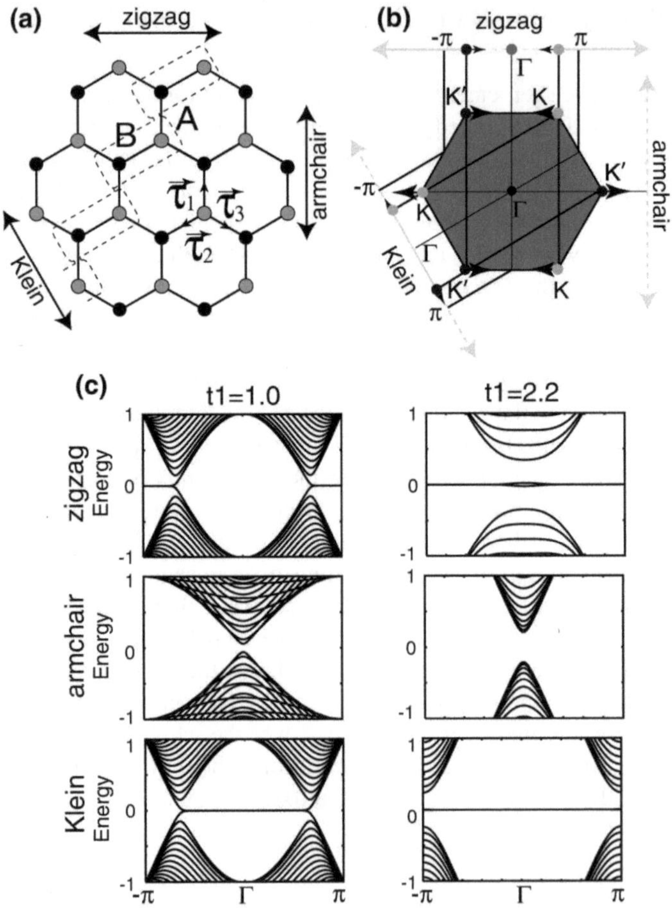

Fig. 2.11 **a** Schematic of the honeycomb-lattice structure. The *dotted line* shows a choice of unit cell with translational symmetry along the edge. The *arrows* show the directions along zigzag, armchair, and Klein edges. **b** The first BZ of the honeycomb lattice. *Dots* at the zone corners show the gap-closing points for the graphene model ($t_1 = t_2 = t_3 = 1$). Namely, the bulk bands become gapless at K and K′ points, when the hopping integral is isotropic. By increasing t_1 from unity, the gap-closing points move away from K and K′ points as shown by the *arrows*. The flat-band edge states expand in the BZ as the bulk gap-closing points move. **c** The dispersions of ribbons for $t_1 = 1$ and 2.2 at $t_2 = t_3 = 1$, with zigzag, armchair and Klein edges. In zigzag and Klein edges, flat bands appear at the zero energy

$$H_{\mathrm{b}}(\mathbf{k}) = \begin{pmatrix} 0 & \sum_{i=1}^{3} t_i e^{-i\mathbf{k}\cdot\tau_i} \\ \sum_{i=1}^{3} t_i e^{i\mathbf{k}\cdot\tau_i} & 0 \end{pmatrix}. \tag{2.72}$$

where the subscript "b" means the bulk. $\boldsymbol{\tau}_{i=1,2,3}$ are expressed as $\boldsymbol{\tau}_1 = (0, 1)$, $\boldsymbol{\tau}_2 = (-\frac{\sqrt{3}}{2}, -\frac{1}{2})$, $\boldsymbol{\tau}_3 = (\frac{\sqrt{3}}{2}, -\frac{1}{2})$, and we put the length of the nearest-neighbor bonds as unity. For simplicity t_i are assumed to be positive. The primitive vectors $\mathbf{a}_{i=1,2}$

are $\mathbf{a}_1 = (\frac{\sqrt{3}}{2}, \frac{3}{2})$, $\mathbf{a}_2 = (-\frac{\sqrt{3}}{2}, \frac{3}{2})$, and the reciprocal primitive vectors are $\mathbf{G}_1 = 2\pi\frac{2}{3}(\frac{\sqrt{3}}{2}, \frac{1}{2})$, $\mathbf{G}_2 = 2\pi\frac{2}{3}(-\frac{\sqrt{3}}{2}, \frac{1}{2})$.

We first note that the bulk Hamiltonian H_b has chiral symmetry:

$$\sigma_z H_\mathrm{b} \sigma_z = -H_\mathrm{b}, \tag{2.73}$$

where σ_z is the Pauli matrix. Therefore, if $|\psi\rangle$ is an eigenstate with an eigenvalue E, $\sigma_z|\psi\rangle$ is an eigenstate with energy $-E$. The eigenvalues are given by

$$E_\mathrm{b}(\mathbf{k}) = \pm\left|\sum_{i=1}^{3} t_i \mathrm{e}^{-i\boldsymbol{\tau}_i\cdot\mathbf{k}}\right|. \tag{2.74}$$

Hereafter we put parameters as $t_2 = t_3 = 1$, and $t_1 = t$, where t is a real positive tunable parameter. The bulk dispersion is given as

$$E_\mathrm{b}^2 = \left(t + 2\cos\frac{\sqrt{3}k_x}{2}\cos\frac{3k_y}{2}\right)^2 + 4\cos^2\frac{\sqrt{3}k_x}{2}\sin^2\frac{3k_y}{2}. \tag{2.75}$$

Because of the chiral symmetry, the gap closes only at zero energy. The bulk gap-closing points (k_x^*, k_y^*) are given by the equations: $\cos\frac{\sqrt{3}k_x^*}{2} = \pm\frac{1}{2}t$ and $\sin\frac{3}{2}k_y^* = 0$. The equations give two gap-closing points in the bulk BZ, and they exist for $t \leq 2$. The gap-closing points move with the change of the anisotropy t, as pointed out by Dietl et al. [58]. For $t = 1$, i.e. the tight-binding model of graphene, the upper and lower bands touch at K $(\frac{2\pi\sqrt{3}}{9}, \frac{2\pi}{3})$ and K′ $(-\frac{2\pi\sqrt{3}}{9}, \frac{2\pi}{3})$, and with the increase of t the gap-closing points get closer along the line $k_y = \frac{2\pi}{3}$ (Fig. 2.11b). Around each of the two gap-closing points, the dispersion forms a Dirac cone, and Berry phase around each gap-closing point is π, which is protected by chiral symmetry. Because of this π Berry phase, the gap-closing points do not disappear as we change $t(< 2)$. The bulk gap-closing points move in the direction perpendicular to the bonds with anisotropic hopping integral t. At $t = 2$ the two points meet and they annihilate each other at $k_x = 0$ (Fig. 2.11b) [58]. This is possible because the sum of the Berry phase becomes zero, i.e. $\pi + \pi \equiv 0 \pmod{2\pi}$. For $t > 2$, there are no bulk gap-closing points.

The evolution of the edge states with the change of the anisotropy has been studied in several papers [58–60]. As we see in the following, for $t > 2$ flat-band edge states on the zigzag or Klein edges completely cover the BZ, as has been studied by Delplace et al. [59]. For the zigzag edges it occurs when the bond with hopping t is perpendicular to the edge, and for the Klein edges it occurs when the bond with hopping t is not perpendicular to the edge. For these cases with zigzag and Klein edges, dispersions are shown in Fig. 2.11c for $t_1 = 1$ and $t_1 = 2.2$, at $t_2 = t_3 = 1$ in both cases. The flat-band edge states are separated completely from the bulk for $t_1 = 2.2$ (Fig. 2.11c).

To explain this behavior, we solve the Schrödinger equation in the semi-infinite geometry with a zigzag edge. The zigzag edge is assumed to be perpendicular to the bonds with hopping integral t_1. We express the wavefunction $|\Psi(k)\rangle$ as

$$|\Psi(k)\rangle = \sum_{i=1}(a_i(k)|A_i(k)\rangle + b_i(k)|B_i(k)\rangle), \tag{2.76}$$

where i denotes an index for unit cells containing two sublattice sites, A and B, counted from the edge ($i = 1$), k is the wavenumber along the edge, and $a_i(k)$ ($b_i(k)$) denotes the coefficient for the wavefunctions at A(B) sublattice, $|A_i(k)\rangle$ ($|B_i(k)\rangle$). Acting H onto $|A_i(k)\rangle$ and $|B_i(k)\rangle$, we have

$$\langle B_i(k)|H|A_i(k)\rangle = t_1, \tag{2.77}$$

$$\langle B_{i-1}(k)|H|A_i(k)\rangle = t_2 + t_3 \mathrm{e}^{-ik}. \tag{2.78}$$

From Fig. 2.11c, the surface states are expected to be at the zero energy, and as we see later it is the case indeed. When we set the eigenvalue to be zero, $H|\Psi(k)\rangle = 0$, we obtain

$$a_i(t_2 + t_3\mathrm{e}^{-ik}) + a_{i+1}t_1 = 0,\ b_i = 0. \tag{2.79}$$

Thus the amplitude of the flat-band states is given by

$$a_n(k) = a_1\left[-\frac{t_2 + t_3\mathrm{e}^{-ik}}{t_1}\right]^{n-1},\ b_i = 0 \tag{2.80}$$

where $a_1 = \left[1 - \frac{|t_2+t_3\mathrm{e}^{-ik}|^2}{t_1^2}\right]^{-1/2}$ from normalization, and the condition for existence of the edge states, i.e. normalizability of the wavefunction, is given as

$$\left|\frac{t_2 + t_3\mathrm{e}^{-ik}}{t_1}\right| < 1. \tag{2.81}$$

For example, for $t_1 = t_2 = t_3$ (graphene model), the wavenumber that satisfies the condition (Eq. 2.81) for existence of the edge state is given as $\frac{2\pi}{3} < k < \frac{4\pi}{3}$, which agrees with the well-known flat band in graphene ribbon with a zigzag edge [55]. In addition, by the relation $\left|\frac{t_2+t_3\mathrm{e}^{-ik}}{t_1}\right| < \frac{t_2+t_3}{t_1}$, when $t_1 > t_2 + t_3$ is satisfied, the wavefunction defined by Eq. (2.80) is normalizable for every $\mathbf{k}$ and the flat bands cover the whole 1D BZ . This condition $t_1 > t_2 + t_3$ means that the anisotropy is sufficiently large. These results agree with numerical calculations in Fig. 2.11c.

In Fig. 2.11c we also show results for armchair edges and for Klein (bearded) edges. For armchair edges, there are no flat-band edge states. For Klein edges where

τ_1-bonds (hopping t_1) are not perpendicular to the edge, there are flat-band edge states; if $t_1 > 2$ the flat-band edge states cover the entire BZ. These results agree with the results by Delplace et al. [59].

2.3.2 Topological Explanation for Existence of the Flat-Band States

In the paper by Ryu and Hatsugai [56], a topological interpretation of the existence of edge states at zero energy in two-dimensional models with chiral symmetry is given. In this section we review this theory by using the tight-binding model on the honeycomb lattice, and show that the flat-band edge states are explained within this theory [56].

To apply the topological argument by Ryu and Hatsugai [56], the crystal termination is crucial. The way how the edges are oriented and how the crystal is terminated is incorporated into the formalism in the following way. For two-dimensional models with chiral symmetry, for example, we begin with a bulk system, and we cut the system along one direction by cutting the nearest-neighbor bonds, in order to discuss edge states. Let y denote the coordinate along which the system will be cut. Then, following the argument by Ryu and Hatsugai [56], we expand the bulk Hamiltonian by the Pauli matrices σ_x, σ_y as $H = h_x(k'_x, k_y)\sigma_x + h_y(k'_x, k_y)\sigma_y$. Here k_y denotes the component of the wavevector along the y-direction (along the edge), and k'_x is the other component of the wavevector. We note that because we assume chiral symmetry $\sigma_z H \sigma_z = -H$, the 2×2 Hamiltonian H has no σ_z term. Because the bulk system is cut along the y-axis, k'_x will no longer be a good quantum number. Then the criterion by Ryu and Hatsugai [56] says that if the trajectory of (h_x, h_y) for the change of k'_x with fixed k_y encircles the origin, zero-energy edge states exist for the given k_y. If not, zero-energy edge states will not exist [56]. An intuitive picture of this argument is the following. The origin $(h_x, h_y) = (0, 0)$ is a singular point because the bulk Hamiltonian has degenerate eigenvalues at zero energy. Whether the trajectory encircles this singularity or not determines a classification of the Hamiltonian either into a class with no edge state or a class with flat-band edge states. Namely, if the trajectory does not encircle the origin, it can be continuously deformed into a point without encountering the singular point, which leads to an absence of zero-energy boundary states.

We apply this criterion to the present models to show that the flat-band boundary states discussed so far are fully explained by this theory. For the anisotropic honeycomb-lattice models, explanations are given by Delplace et al. [59], and we reproduce it here for illustration. For zigzag edges we have

$$h_x = t\cos(k_y - k'_x) + 1 + \cos k_y, \tag{2.82}$$

$$h_y = -t\sin(k_y - k'_x) + \sin k_y. \tag{2.83}$$

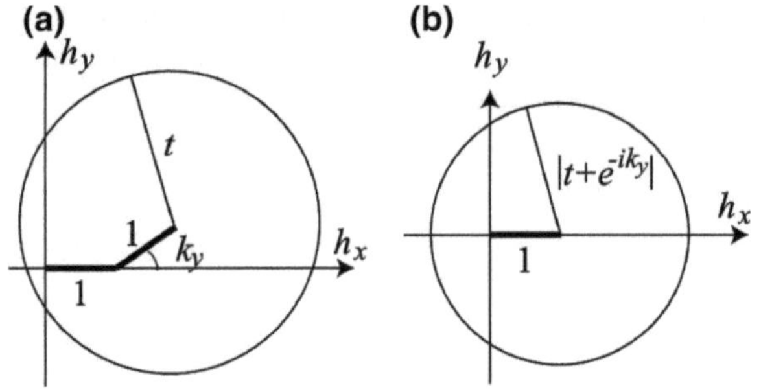

Fig. 2.12 Trajectories of (h_x, h_y) by varying k_x' in (**a**, **b**) for the honeycomb-lattice model with zigzag (**a**), with Klein (**b**) edges

Hence the trajectory is a circle with a radius $|t|$ centered at $(1 + \cos k_y, \sin k_y)$ (Fig. 2.12a). The condition that it encircles the origin reproduces the range of the wavevector of the flat-band edge states, obtained in the previous section. In particular, for $t > 2$ the trajectory encompasses the origin irrespective of the value of k_y, and existence of the perfectly flat edge band over the whole BZ results, as we discussed previously. The case for the Klein edge is explained similarly, where we have

$$h_x = 1 + \cos(k_y - k_x') + t \cos k_x', \tag{2.84}$$

$$h_y = -\sin(k_y - k_x') + t \sin k_x'. \tag{2.85}$$

with the trajectory shown in Fig. 2.12b. Then it is easily seen that the flat-band edge states extend over the whole BZ when $t > 2$.

2.3.3 Completely Localized Edge States

We have shown that the nearest-neighbor tight-binding models on the honeycomb lattice. have flat-band boundary states covering the whole BZ, when the anisotropy of their hopping integrals is sufficiently large. In general, when systems have completely flat bands over the entire BZ, one can construct a wavefunction which is spatially localized on a finite number of sites. Namely, because of the flatness of the band, any linear combination of the eigenstates within this flat band is also an eigenstate; therefore by taking an appropriate linear combination, one can construct a fully localized state. This is analogous to constructing a spatially localized state as a linear combination of plane waves.

The construction of the fully localized state is possible only when the flat band covers the whole BZ [57]. In this section, we calculate the fully localized wavefunction in the present models. This wavefunction is exponentially decaying in the direction normal to the boundary, while on the outermost atomic layer, the wavefunction is nonzero only on a single site, as schematically shown in Fig. 2.13. We consider semi-infinite systems for the honeycomb lattice with the zigzag edge.

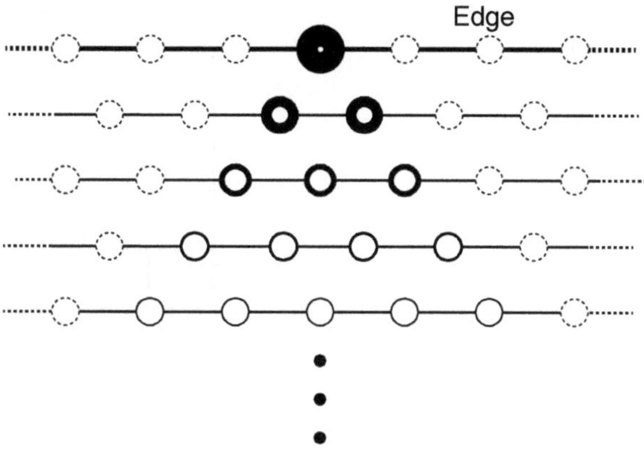

Fig. 2.13 Schematic of spatial distribution of the flat-band surface states. The *circles* are the A sublattice sites, while the amplitudes on B sublattices are zero and are omitted. The *dotted circles* show that their amplitudes of the wavefunction are zero. The *line thickness* of the *circle* shows the magnitude of the amplitude. The *top* atom has the largest amplitude, and the distribution of the wavefunction spatially spreads toward the interior with exponential decay

Similarly to the previous sections, the outermost atomic layer is assumed to belong to the A sublattice.

We consider the honeycomb lattice in the half plane $y \leq 0$. The zigzag edge is along the x axis, and the origin $\mathcal{O}$ is set to be one site on the zigzag edge. The state at the site $-m\mathbf{a}_1 - n\mathbf{a}_2$ is denoted as $|A_{mn}\rangle$ where $\mathbf{m} = (m, n)$ are nonnegative integers, and $\mathcal{O} = (0, 0)$. We assume that the amplitude of the localized wavefunction on the outermost atomic layer is nonzero only at $\mathcal{O}$. We express the wavefunction of the localized states as

$$|\Psi\rangle = \sum_{m,n} a_{mn} |A_{mn}\rangle, \tag{2.86}$$

where a_{mn} is the amplitude for $|A_{mn}\rangle$ at A sublattice. The amplitude at B sublattice is identically zero. From the Schrödinger equation, we have a relation between amplitudes as

$$a_{mn} = -\frac{t_2}{t_1} a_{m-1n} - \frac{t_3}{t_1} a_{mn-1}. \tag{2.87}$$

Intriguingly, the solution for the above sequence determined by (2.87) is the same as the following problem. Consider a mover in the xy plane on the grid shown in Fig. 2.14. The mover is first on the $\mathcal{O} = (0, 0)$ site. At each step it moves by $(1, 0)$ with a probability P_1, by $(0, 1)$ with a probability P_2 (Fig. 2.14), and the movement is finished otherwise. Finally after $m + n$ steps the probability P_{mn} that the mover is at (m, n) $(m, n \geq 0)$ along the shortest paths is given as

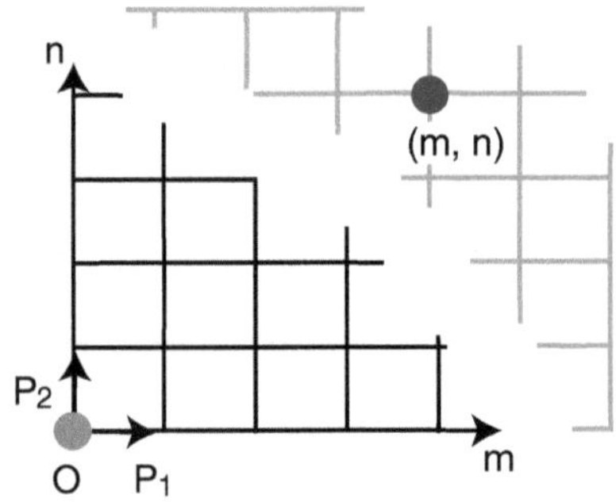

Fig. 2.14 Schematic of the grids for the shortest path problem. The mover is at first at the origin $\mathcal{O}$, and it moves along $\mathbf{e}_i$ with the probability P_i, where $(\mathbf{e}_i)_j = \delta_{ij}$

$$P_{mn} = \frac{(m+n)!}{m!n!} P_1^m P_2^n. \tag{2.88}$$

By replacing the probabilities of the movement with the ratio of the hopping integrals,

$$P_i \to -\frac{t_{i+1}}{t_1}, \tag{2.89}$$

where $i = 1, 2$, we have the amplitude of the wavefunction as

$$a_{mn} = \frac{1}{Z}\frac{(m+n)!}{m!n!}\left(-\frac{t_2}{t_1}\right)^m\left(-\frac{t_3}{t_1}\right)^n \quad (m, n \geq 0), \tag{2.90}$$

where Z is the normalization constant. For $m < 0$ or $n < 0$, a_{mn} vanishes. In this spatial representation of the wavefunction, the condition for existence of the fully localized states on the boundary is that the wavefunction is normalizable. Generally, because $t_{1,2,3}$ are positive, Z satisfies the following relation:

$$\begin{aligned} Z^2 &= \sum_{N=0}^{\infty}\sum_{n,m=0}^{N} \delta_{n+m,N} P_{mn}^2 \\ &\leq \left(\sum_{N=0}^{\infty}\sum_{n,m=0}^{N=n+m} |P_{mn}|\right)^2 = \left[\sum_{N=0}^{\infty}\left(\frac{t_2+t_3}{t_1}\right)^N\right]^2. \end{aligned} \tag{2.91}$$

Therefore for $\frac{t_2+t_3}{t_1} < 1$, the normalization constant Z converges; namely the fully localized states appear. This condition is the same as that for the appearance of flat-band edge states.

References

1. K. von Klitzing, K.G. Dorda, M. Pepper, Phys. Rev. Lett. **45**, 494 (1980)
2. M.Z. Hasan, C.L. Kane, Rev. Mod. Phys. **82**, 3045 (2010)
3. R.B. Laughlin, Phys. Rev. B **23**, 5632 (1981)
4. D.J. Thouless, M. Kohmoto, M.P. Nightingale, M. den Nijs, Phys. Rev. Lett. **49**, 405 (1982)
5. M.V. Berry, Proc. R. Soc. Lond. A **392**, 45 (1984)
6. J.E. Avron, D. Osadchy, R. Seiler, Phys. Today **56**, 38 (2003)
7. M. Kohmoto, Ann. Phys. N.Y. **160**, 343 (1985)
8. R.D. King-Smith, D. Vanderbilt, Phys. Rev. B **47**, 1651–1654 (1993)
9. R. Resta, Rev. Mod. Phys. **66**, 899–915 (1994)
10. B.I. Halperin, Phys. Rev. B **25**, 2724 (1982)
11. M. Büttiker, Phys. Rev. B **38**, 9375 (1988)
12. F.D.M. Haldane, Phys. Rev. Lett. **61**, 2015 (1988)
13. C.L. Kane, E.J. Mele, Phys. Rev. Lett. **95**, 226801 (2005)
14. C.L. Kane, E.J. Mele, Phys. Rev. Lett. **95**, 146802 (2005)
15. B.A. Bernevig, T.L. Hughes, S.-C. Zhang, Science **314**, 1757 (2006)
16. J.E. Moore, L. Balents, Phys. Rev. B **75**, 121306(R) (2007)
17. L. Fu, C.L. Kane, E.J. Mele, Phys. Rev. Lett. **98**, 106803 (2007)
18. R. Roy, Phys. Rev. B **79**, 195322 (2009)
19. M. König et al., Science **318**, 766 (2007)
20. R. Roy, Phys. Rev. B **79**, 195321 (2009)
21. D. Hsieh et al., Nature (London) **452**, 970 (2008)
22. Y. Xia et al., Nat. Phys. **5**, 398 (2009)
23. D. Hsieh et al., Nature (London) **460**, 1101 (2009)
24. Y.L. Chen et al., Science **325**, 178 (2009)
25. T. Valla et al., Phys. Rev. B **86**, 241101(R) (2012)
26. C. Xu, J.E. Moore, Phys. Rev. B **73**, 045322 (2006)
27. C. Wu, B.A. Bernevig, S.-C. Zhang, Phys. Rev. Lett. **96**, 106401 (2006)
28. H.J. Goldsmid, *Thermoelectric Refrigeration* (Plenum, New York, 1964)
29. G.A. Slack, in *CRC Handbook of Thermoelectrics* ed by D.M. Rowe (CRC Press, Boca Raton, 1995), pp. 407–440
30. R. Takahashi, S. Murakami, Phys. Rev. B **81**, 161302(R) (2010)
31. P. Ghaemi, R.S.K. Mong, J.E. Moore. Phys. Rev. Lett. **105**, 166603 (2010)
32. L.D. Hicks, M.S. Dresselhaus, Phys. Rev. B **47**, 12727 (1993)
33. L.D. Hicks, M.S. Dresselhaus, Phys. Rev. B **47**, 16631 (1993)
34. R. Takahashi, S. Murakami, Semicond. Sci. Technol. **27**, 124005 (2012)
35. L. Fu, C.L. Kane, Phys. Rev. B **74**, 195312 (2006)
36. S. Murakami, N. Nagaosa, S.-C. Zhang, Science **301**, 1348 (2006)
37. J. Sinova et al., Phys. Rev. Lett. **92**, 126603 (2004)
38. M.I. Dyakanov, V.I. Perel, JETP Lett. **13**, 467 (1971)
39. M.I. Dyakanov, V.I. Perel, Phys. Lett. A **35**, 459 (1971)
40. Y.K. Kato et al., Science **306**, 1910 (2004)
41. J. Wunderlich et al., Phys. Rev. Lett. **94**, 047204 (2005)
42. X.L. Qi et al., Phys. Rev. B **78**, 195424 (2008)
43. Z. Wang et al., New J. Phys. **12**, 065007 (2010)
44. A.M. Essin, J.E. Moore, D. Vanderbilt, Phys. Rev. Lett. **102**, 146805 (2009)
45. L. Fu, C.L. Kane, Phys. Rev. B **76**, 045302 (2007)
46. H. Zhang et al., Nat. Phys. **5**, 438 (2009)
47. C.X. Liu et al., Phys. Rev. B **82**, 045122 (2010)
48. W.-Y. Shan, H.-Z. Lu, S.-Q. Shen, New J. Phys. **12**, 043048 (2010)
49. E.H. Lieb, Phys. Rev. Lett. **62**, 1201 (1989)
50. A. Mielke, J. Phys. A: Math. Gen. **24**, L73 (1991)
51. A. Mielke, J. Phys. A **24**, 3311 (1991)

52. A. Mielke, J. Phys. A **25**, 4335 (1992)
53. A. Mielke, H. Tasaki, Commun. Math. Phys. **158**, 341 (1993)
54. K.S. Novoselov et al., Science **306**, 666 (2004)
55. M. Fujita et al., J. Phys. Soc. **65**, 1920 (1996)
56. S. Ryu, Y. Hatsugai, Phys. Rev. Lett. **89**, 077002 (2002)
57. R. Takahashi, S. Murakami, Phys. Rev. B **88**, 235303 (2013)
58. P. Dietl, F. Piechon, G. Montambaux, Phys. Rev. Lett. **100**, 236405 (2008)
59. P. Delplace, D. Ullmo, G. Montambaux, Phys. Rev. B **84**, 195452 (2011)
60. M. Kohmoto, Y. Hasegawa, Phys. Rev. B **76**, 205402 (2007)

Chapter 3
Gapless Interface States Between Two Topological Insulators

3.1 Refractive Phenomena of the Surface States at the Boundary of Two Topological Insulators

Recently physical phenomena originating from the Dirac cones of electrons have been studied, in the context of graphene sheet [1–4] and the topological insulators (TI) [5, 7–9]. In a graphene sheet, novel transport phenomena are predicted theoretically in *p*-*n* junction systems: for example the Klein paradox [10], and the negative refraction [11]. The TI in three dimension (3D) [8, 9], such as Bi_2Se_3 [12, 13] and Bi_2Te_3 [14], has a single Dirac cone in its surface states, as observed in angle-resolved photoemission spectroscopy. Unlike graphene, the states on the Dirac cone on the surface of the TI are spin-filtered; they have a fixed spin direction for each wavenumber **k**. Because two states at **k** and −**k** on the Dirac cone have the opposite spins, perfect backscattering from **k** to −**k** is forbidden. In addition, the velocity of the Dirac cone on the surface of 3DTI depends on materials. For example, the velocity for Bi_2Te_3 is about 4×10^5 m/s [14] depending on the direction of the wave vector, and that for Bi_2Se_3 is approximately 5×10^5 m/s [12]. Therefore, when two different TIs are attached together, a refraction phenomenon similar to optics is expected at the junction. In this section, we study refraction of electrons at the junction between the surfaces of two TIs (Fig. 3.1). The resulting transmittance and reflectance are different from optics, and they reflect prohibited perfect backscattering.

The effective Dirac Hamiltonian of the surface states on the xz-plane (Fig. 3.1) is represented as

$$H = -iv[\sigma_x \partial_z - \sigma_y \partial_x], \tag{3.1}$$

where σ_x, σ_y are the Pauli matrices, and v is the Fermi velocity. Here the spin quantization axis is set to be the y-axis. The eigenstate $\boldsymbol{\psi}(\mathbf{k})$ and eigenvalue E are

R. Takahashi, *Topological States on Interfaces Protected by Symmetry*, Springer Theses, DOI 10.1007/978-4-431-55534-6_3

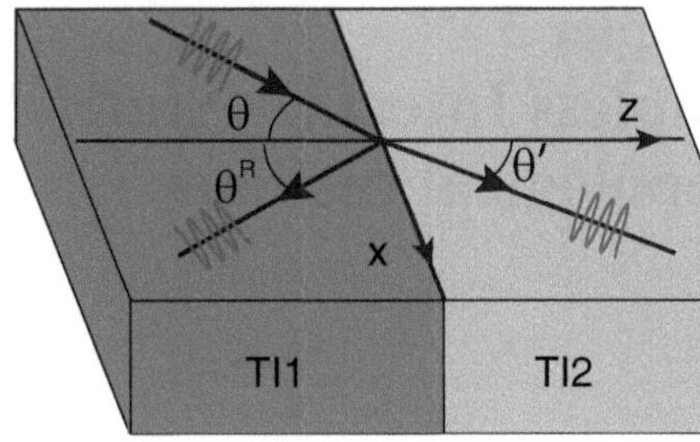

Fig. 3.1 Schematic of the refraction of the surface states at the junction between the two TIs, TI1 and TI2. In TI1, the incidence angle is θ, and the reflection angle is $\theta_R = \theta$. In TI2 the transmission angle is θ'

$$\psi(\mathbf{k}) = \frac{1}{\sqrt{2}} \begin{pmatrix} 1 \\ s\mathrm{e}^{-i\phi_{\mathbf{k}}} \end{pmatrix} \mathrm{e}^{i\mathbf{k}\cdot\mathbf{r}}, \tag{3.2}$$

$$E = svk, \tag{3.3}$$

where k is the norm of the wavevector $\mathbf{k} = (k_x, k_z)$, $s = +1(-1)$ corresponds to the upper (lower) Dirac cone, provided $v > 0$, and $\phi_{\mathbf{k}} = \arctan(k_x/k_z)$. We consider a refraction problem between two TIs, which we call TI1 and TI2, with the incidence angle θ, the transmission angle θ', and the reflection angle θ^R (Fig. 3.1). As in optics, the momentum conservation requires $\theta^R = \theta$, and the wave functions are written as

$$\psi^I(x, z) = \frac{1}{\sqrt{2}} \mathrm{e}^{ik(x\sin\theta + z\cos\theta)} \begin{pmatrix} 1 \\ \mathrm{e}^{-i\theta} \end{pmatrix}, \tag{3.4}$$

$$\psi^T(x, z) = \frac{1}{\sqrt{2}} \mathrm{e}^{ik'(x\sin\theta' + z\cos\theta')} \begin{pmatrix} 1 \\ \mathrm{e}^{-i\theta'} \end{pmatrix}, \tag{3.5}$$

$$\psi^R(x, z) = \frac{1}{\sqrt{2}} \mathrm{e}^{ik(x\sin\theta - z\cos\theta)} \begin{pmatrix} 1 \\ -\mathrm{e}^{i\theta} \end{pmatrix}, \tag{3.6}$$

where k and k' are the wavenumbers on TI1 and TI2 surfaces, respectively, and we consider the Fermi energy $E_F > 0$ (i.e. above the Dirac point), giving $s = +1$ for both of the TIs. Let v_1 and v_2 denote the velocities of the two TIs surface states.

We first assume v_1 and v_2 to be positive. Due to the conservation of the momentum and the energy, Snell's law appears as

$$k' \sin\theta' = k\sin\theta, \ v_1^{-1}\sin\theta = v_2^{-1}\sin\theta'. \tag{3.7}$$

Let r and t denote the amplitudes of the reflected and transmitted waves, compared with the incident wave. The current conservation in the z direction is written as

$$1 - |r|^2 = \frac{v_2\cos\theta'}{v_1\cos\theta}|t|^2. \tag{3.8}$$

In the present problem, the ordinary boundary condition of the continuity of the wave functions at the boundary, $\psi^I + r\psi^R = t\psi^T$, leads to a contradiction; namely this boundary condition gives $r = i\frac{\sin\frac{\theta'-\theta}{2}}{\cos\frac{\theta+\theta'}{2}}e^{-i\theta}$, $t = e^{i\frac{(\theta'-\theta)}{2}}\frac{\cos\theta}{\cos\frac{(\theta'+\theta)}{2}}$, which do not satisfy Eq. (3.8). We note that the wavefunction should eventually be discontinuous at the junction when the velocities are different. The particle current is given by $|\psi|^2\mathbf{v}$, where $\mathbf{v}$ is a velocity vector. Therefore, the current conservation at the interface between the TIs with different velocities requires $v_{1z}|\psi_1|^2 = v_{2z}|\psi_2|^2$. Because in our case $v_{1z} \neq v_{2z}$, we have $|\psi_1|^2 \neq |\psi_2|^2$ at the junction, and the continuity of the wavefunction is violated. This is related with the hermiticity of the Hamiltonian. The proper way to solve the problem is to set the Hamiltonian to be hermitian also at the boundary. Because the velocity depends on z, to remain hermitian the effective Hamiltonian should read

$$H = -i\left[\frac{1}{2}[v(z)\sigma_x\partial_z + \sigma_x\partial_z v(z)] - v(z)\sigma_y\partial_x\right]. \tag{3.9}$$

Furthermore, to avoid the discontinuity of v, we temporarily assume v to be the following function (Fig. 3.2):

$$v(z) = \begin{cases} v_1 & z < -\frac{a}{2} \\ \frac{v_2-v_1}{a}z + \frac{v_1+v_2}{2} & -\frac{a}{2} < z < \frac{a}{2} \\ v_2 & \frac{a}{2} < z \end{cases}, \tag{3.10}$$

where a is the width of the boundary region whose center is at $z = 0$. Because this $v(z)$ is continuous, we can impose continuity for the wavefunction at $z = \pm\frac{a}{2}$ This parameter a will be set to zero at the end of the calculation.

The Schrödinger equation in the boundary region $-\frac{a}{2} < z < \frac{a}{2}$ has the form

$$\begin{cases} E\psi_1 = i(vk_x - v\partial_z - \frac{\Delta v}{2a})\psi_2 \\ E\psi_2 = -i(vk_x + v\partial_z + \frac{\Delta v}{2a})\psi_1 \end{cases}, \tag{3.11}$$

where $\Delta v = v_2 - v_1$, E is the eigenvalue, k_x is the wavevector due to the translational symmetry for the x axis, and $\psi(z) = [\psi_1(z), \psi_2(z)]^t$ is the wavefunction. When we change the variable $z \to v(z)$ and bring together the above equations, we obtain the differential equation

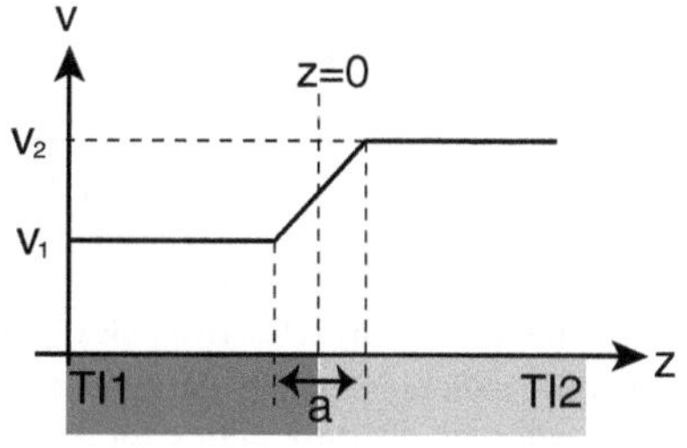

Fig. 3.2 Dirac velocity v in our model as the function of z (Eq. 3.10) for $v_1 < v_2$. To avoid discontinuity of wavefunctionns at the junction, the velocity is taken to be of this form at first, and a will be set to be zero at the end of the calculation

$$\partial_v(v^2\psi_1) - (\alpha^2 v^2 - \alpha v - \frac{1}{4} - \varepsilon^2)\psi_1 = 0, \tag{3.12}$$

where $\alpha = \frac{ak_x}{\Delta v}$, $\varepsilon = \frac{aE}{\Delta v}$, and we assume $v_{1(2)} < v < v_{2(1)}$. After solving the differential equation, we take $a \to 0$, because the width of the boundary is assumed to be very small compared with the system size. We impose the continuity of the wavefunctions, $\psi(v_2) = \psi^T\left(\frac{a}{2}\right)$ at $z = \frac{a}{2}$ and $\psi(v_1) = C\psi^I\left(-\frac{a}{2}\right) + D\psi^R\left(-\frac{a}{2}\right)$ at $z = -\frac{a}{2}$ where C and D are the coefficients determined by the boundary condition. The resulting coefficients are

$$r = i\frac{\sin\frac{\theta'-\theta}{2}}{\cos\frac{\theta+\theta'}{2}}e^{-i\theta}, \quad t = \sqrt{\frac{v_1}{v_2}}\frac{\cos\theta}{\cos\frac{\theta+\theta'}{2}}e^{i\frac{\theta'-\theta}{2}}. \tag{3.13}$$

They satisfy the current conservation Eq. (3.8), and make the wavefunction discontinuous in the limit $a \to 0$. Equation (3.13) gives the transmittance T and the reflectance R as

$$R = |r|^2 = \frac{\sin^2\frac{\theta'-\theta}{2}}{\cos^2\frac{\theta+\theta'}{2}}, \tag{3.14}$$

$$T = \frac{v_2}{v_1}\frac{\cos\theta'}{\cos\theta}|t|^2 = \frac{\cos\theta\cos\theta'}{\cos^2\frac{\theta+\theta'}{2}} = 1 - R. \tag{3.15}$$

The results are plotted as solid curves in Fig. 3.3a, b. The dotted curves represent corresponding results for the optic refractive phenomena.[1] Unlike optics, for normal

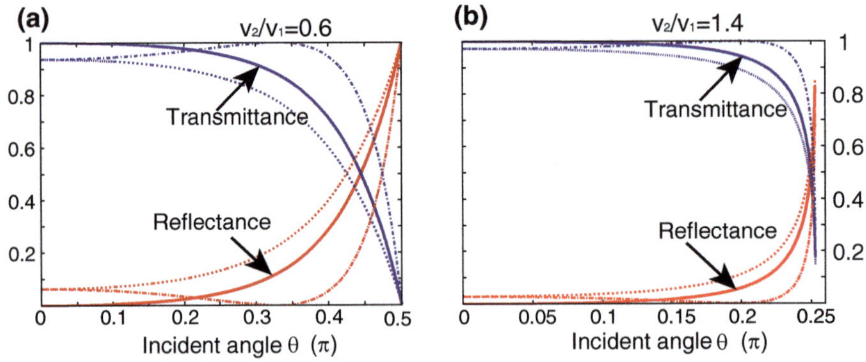

Fig. 3.3 a, b Reflectance (*red*) and transmittance (*blue*) for the ratios of the velocities of the two TIs: **a** $v_2/v_1 = 0.6$ and **b** $v_2/v_1 = 1.4$. The *solid curves* are the results for the junction between two TIs, while the *dotted curves* show the results for optics with p- and s-polarizations

[1] From Eq. (3.7), the refractive index of TI1(2) $n_{1(2)}$ is proportional to $v_{1(2)}^{-1}$. Then the ratio of the refractive indices is equal to the inverse of the ratio of the velocities, $\frac{n_1}{n_2} = \frac{v_2}{v_1}$; thereby we compare the results with those for optics, for the same ratios between the two velocities.

incidence ($\theta = 0$), we have $T = 1$ and $R = 0$; namely the perfect transmission occurs when the incident wave is perpendicular to the interface, which reflects the prohibited backscattering on the surface of the TI. This is similar to graphene [10] but the transmittance in our case monotonically decreases with the incidence angle.

3.2 Gapless Interface States Between Two TIs

3.2.1 *Paradox in the Refractive Phenomena*

Next, we consider the case where the velocities of the two TIs have opposite signs. One might think that it is similar to the negative refraction in optics [15, 16], but it is not true because the Fermi energy is above the Dirac point for the two TIs. In the present case, the aforementioned approach is no longer useful, because the method of putting the velocity continuous (Eq. 3.10) leads to a vanishing velocity somewhere. Furthermore, both reflection and transmission for the surface transport are prohibited for normal incidence, because the incident wave has no way to conserve its momentum and spin simultaneously (see Fig. 3.4). Thus it is a paradox what happens when the incident wave is perpendicular to the junction.

Our answer to this question is the following. Here we show existence of gapless states at the interface between the two TIs (the purple region in Fig. 3.4b). The normally incident wave goes along the surface of one TI, then into the interface between the two TIs. These interface states arise from hybridization between the two surface states from the two TIs.

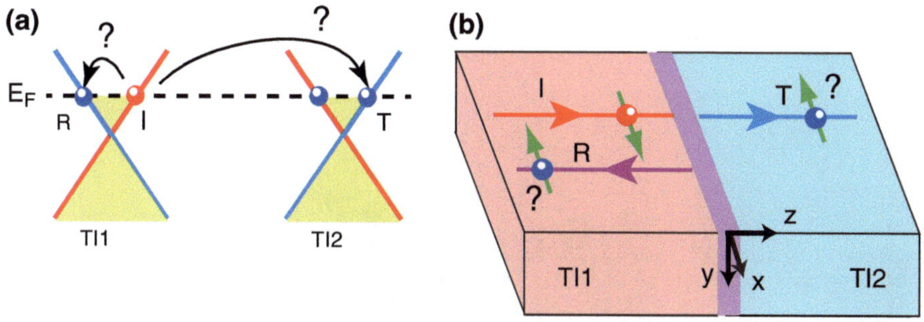

Fig. 3.4 Schematics of dispersions and the transport on the surface states on the two TI whose surface velocities have different signs. **a** Linear dispersion at $k_x = 0$. The incident wave ("I") is perpendicular to the junction. Both the transmission ("T") and reflection ("R") are prohibited due to spin conservation. **b** Illustration for normal incidence. TI1 (*red*) and TI2 (*blue*) have the velocities of opposite signs

3.2.2 Gapless States from the Surface Effective Model

To show the existence of the gapless interface states, we first write down an effective Hamiltonian at the interface (3.5) from the two Dirac cones with hybridization:

$$H = \begin{pmatrix} H_1 & V \\ V^\dagger & H_2 \end{pmatrix}. \tag{3.16}$$

Here $H_{1(2)}$ is the effective surface Hamiltonian for the surface of TI1 (TI2), constituting the interface:

$$\begin{cases} H_1 = v_1(\boldsymbol{\sigma} \times \mathbf{k})_z, \\ H_2 = -v_2(\boldsymbol{\sigma} \times \mathbf{k})_z, \end{cases} \tag{3.17}$$

and V is the hybridization at the interface. For simplicity, we retain only the lowest order in $\mathbf{k}$. In the expression of H_2, there is an extra minus sign, because the surface normal vectors for TI1 and TI2 are opposite to each other (Fig. 3.5). We restrict the form of the hybridization V by imposing the symmetries of the system. First, we impose time-reversal symmetry. $\Theta V \Theta^{-1} = V$ where Θ is the time-reversal operator $\Theta = i\sigma_y K$, and K is the complex conjugation operator. Then, V is expressed as

$$V = \begin{pmatrix} \tilde{g} & \tilde{h} \\ -\tilde{h}^* & \tilde{g}^* \end{pmatrix}, \tag{3.18}$$

where $\tilde{g}$ and $\tilde{h}$ are complex constants representing the hybridization between the two surface states. From the Hamiltonian (Eq. 3.16), the eigenvalues are calculated as

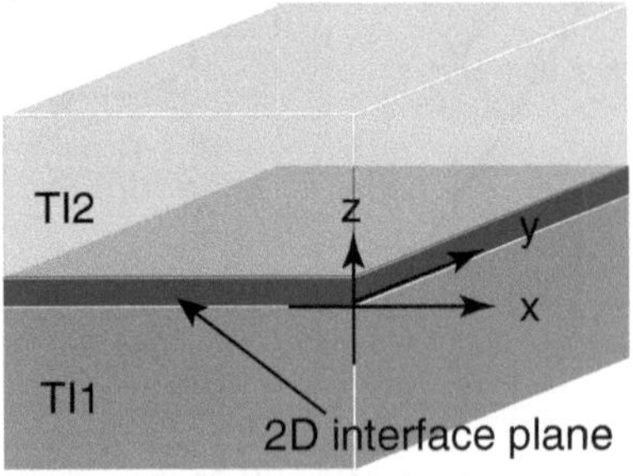

Fig. 3.5 Schematic of the interface plane. We consider the two-dimensional (2D) interface plane between the two TIs. In the low-energy model (Eq. 3.16), each surface facing the interface contributes one Dirac cone to the interface Hamiltonian, and we also add the hybridization V between the wavefunctions of the two TIs

$$E = \pm\sqrt{\Delta_k \pm \sqrt{\Delta_k^2 - \eta}}, \tag{3.19}$$

$$\eta = v_1^2 v_2^2 k^4 + \Delta_0^2 + 2v_1 v_2 k^2 [h^2 \cos(2\alpha + 2\gamma) + g^2 \cos(2\beta)], \tag{3.20}$$

$$\Delta_k = g^2 + h^2 + \frac{v_1^2 + v_2^2}{2} k^2, \tag{3.21}$$

where we set $k_x + ik_y = ke^{i\alpha}$, $\tilde{g} = ge^{i\beta}$, $\tilde{h} = he^{i\gamma}$, and $k, g, h, \alpha, \beta, \gamma \in \Re$. The condition for existence of gapless interface states is

$$v_1^2 v_2^2 k^4 + 2v_1 v_2 k^2 [h^2 \cos(2\alpha + 2\gamma) + g^2 \cos(2\beta)] + \Delta_0^2 = 0.$$

This leads to

$$\begin{cases} \cos(2\alpha + 2\gamma) = \cos(2\beta) = -1 \quad (v_1 v_2 > 0), \\ \text{or} \\ \cos(2\alpha + 2\gamma) = \cos(2\beta) = 1 \quad (v_1 v_2 < 0), \end{cases} \tag{3.22}$$

and the wavenumbers where the gap closes are

$$k = \pm\sqrt{\frac{\Delta_0}{|v_1 v_2|}}. \tag{3.23}$$

In addition, if we impose a mirror symmetry with respect to the yz plane $\mathcal{M}_{yz} = -i\sigma_x P$ where P is the inversion operator, the form of the hybridization V is further restricted. This symmetry is natural in the present problem because the mirror symmetry fixes the spins in plane normal to the yz plane, and this coincides with the situation in Fig. 3.4b where the electrons moving in yz plane have the spin along the x direction. Hence, V is restricted as

$$\mathcal{M}_{yz} V \mathcal{M}_{yz}^{-1} = V, \tag{3.24}$$

and then we have

$$\tilde{g} = \tilde{g}^*, \ \tilde{h} = -\tilde{h}^*. \tag{3.25}$$

Namely $\tilde{g}$ is real, $\beta = 0$, $\tilde{h}$ is pure imaginary $\tilde{h} = ih$, and $\alpha = \pm\frac{\pi}{2}$. Therefore, to obtain the gapless states, we choose the latter condition of Eq. (3.22). Then we see that for $v_1 v_2 > 0$ (the two velocities with the same signs), the interface states are gapped by the hybridization. Figure 3.6 shows the dispersions by Eq. (3.19). In Fig. 3.6a, b, the parameters are $g = 2.0$ and $h = 1.0$, with the Dirac velocities being $v_1 = 1.0$, $v_2 = -2.0$ in Fig. 3.6a and $v_1 = 1.0$, $v_2 = 2.0$ in Fig. 3.6b. In Fig. 3.6a, there are two Dirac cones, and in Fig. 3.6b there is no Dirac cone. The result agrees with the above discussion.

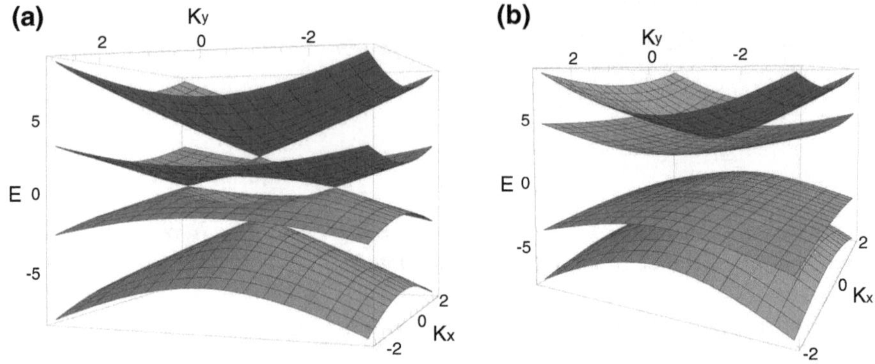

Fig. 3.6 Dispersions on the interface between the two TIs in Eq. (3.19) with velocities $v_1 = 1.0$, $v_2 = -2.0$ (**a**), and with $v_1 = 1.0$, $v_2 = 2.0$ (**b**). The parameters of the hybridization are employed as $g = 2.0$ and $h = 1.0$ in both results. Because E_F is assumed to be zero, only the result **a** has the interface states for $v_1 v_2 < 0$

Only when the two velocities have opposite signs ($v_1 v_2 < 0$), are there gapless states on the interface by the above discussion. Dispersion of the gapless states depends on h. When $h \neq 0$, Eq. (3.22) has solutions only when $v_1 v_2 < 0$, and they are

$$(k_x, k_y) = \left(0, \pm\sqrt{\frac{g^2 + h^2}{|v_1 v_2|}}\right). \tag{3.26}$$

The interface states have two Dirac cones (Fig. 3.6a). We note that g is nonzero in general, whereas h becomes zero when additional symmetries such as rotational symmetry with respect to the z axis are imposed. When h becomes zero due to rotational symmetry with respect to the z axis, the gap-closing points form a circle (Fig. 3.7a)

$$k_x^2 + k_y^2 = \frac{g^2}{|v_1 v_2|}. \tag{3.27}$$

This degeneracy along the circle in **k** space is due to the continuous rotational symmetry of the model Hamiltonian around the z axis, and is lifted when it is broken by adding higher-order terms in **k**, which is realized e.g. by adding the warping terms in Bi_2Se_3 [17] (Fig. 3.7b). We emphasize that the gapless points (Eq. 3.26) exist irrespective of the magnitude of hybridization. Therefore, for this model Hamiltonian, we have shown that there are gapless states at the interface when the system has the mirror symmetry $\mathscr{M}_{yz}$.

From the above results for the interface states, we now have the answer to the paradox in the first part of Sect. 3.2.1. As shown in Fig. 3.8, the incident waves on the

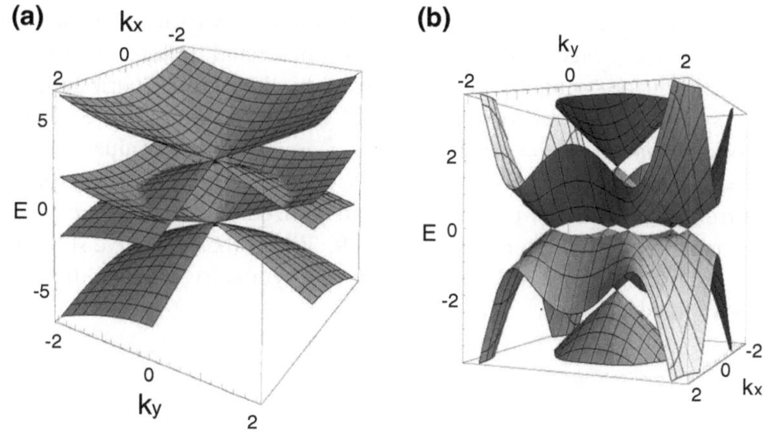

Fig. 3.7 Dispersion of the interface states between the two TIs in Eq. (3.19) with parameters $v_1 = 1.0$, $v_2 = -2.0$ and $g = 2.0$, $h = 0.0$. In (**a**) the region $k_x > 0, k_y > 0$ is omitted for clarity. The gap closes along a *circle* in the k space. Because the system has the continuous rotational symmetry, the omitted dispersion can be complemented by rotation. In (**b**), the warping term $\lambda(k_+^3 + k_-^3)\sigma_z$ with $\lambda = 0.4$ is added to H_1 and H_2. There appear six Dirac cones

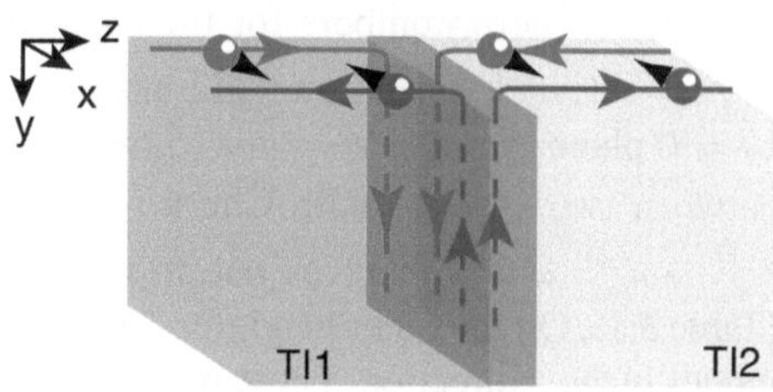

Fig. 3.8 Illustration of the interface states between two TIs with opposite Dirac velocities. The surface current goes into the interface

surface all go into the gapless interface states, with neither refraction nor transmission to the surface states. We note that we can generalize the argument for the existence of the gapless interface states. To give a general proof, we introduce a topological number called the mirror Chern number in the next section.

3.2.3 Proof for the Existence of the Gapless States by the Mirror Chern Number

In the previous section, we show the dispersion of the gapless states by using the effective low-energy model at the interface between two TIs with opposite signs of the Dirac velocities. The existence of gapless interface states is not a coincidence, but can be shown on generic grounds. Because these gapless states are generated between two TIs with the same Z_2 topological numbers, they are not protected in the same sense as the surface states of 3D TIs. In the following we show that these gapless interface states are protected by the mirror symmetry and time-reversal symmetry. Each TI with mirror symmetry is characterized by the mirror Chern number [18]. When the system has the mirror symmetry $\mathscr{M}_{yz}$, the surface modes are labeled with the mirror eigenvalues $\mathscr{M} = \pm i$. Because $\mathscr{M}^2 = -1$ the eigenvalues are $\pm i$. Due to $\mathscr{M}_{yz} = -i\sigma_x P$, $\mathscr{M} = \pm i$ corresponds to $S_x = \mp$, i.e. the spins are perpendicular to the mirror plane. These two subspaces characterized by $\mathscr{M} = \pm i$ are decoupled, and one can define the Chern numbers $n_{\pm i}$ [19, 20] for the respective subspaces. In the present case, the states at $k_x = 0$ satisfy the symmetry, and the mirror Chern number is defined as

$$n_{\mathscr{M}} = \frac{n_{+i} - n_{-i}}{2}. \tag{3.28}$$

We have $n_{+i} = -n_{-i}$ by the time reversal symmetry, and therefore $n_{\mathscr{M}} = n_i = -n_{-i}$. If the mirror Chern number is zero under the time-reversal symmetry, the two Chern numbers are also zero.

In the QH system, the sign of the Chern number gives the direction of the chiral edge modes, either clockwise or counterclockwise. In a similar sense, in the present case where the two Dirac cones from the two TI surfaces have opposite velocities, the mirror Chern numbers for the two TIs are different. TI1 has $n^{(1)}_{\mathscr{M}} = -1$, i.e. $n^{(1)}_{+i} = -1, n^{(1)}_{-i} = +1$, and TI2 has $n^{(2)}_{\mathscr{M}} = 1$, i.e. $n^{(2)}_{+i} = +1, n^{(2)}_{-i} = -1$ on the $k_x = 0$ plane. For the $\mathscr{M} = +i$ $(S_x < 0)$ subspace, this corresponds to the junction between two systems with Chern numbers $n^{(1)}_{+i} = -1$ and $n^{(2)}_{+i} = +1$; because $n^{(1)}_{+i} - n^{(2)}_{+i} = -2$, it gives rise to two left-going chiral modes in the y direction (Table 3.1). On the other hand, for $\mathscr{M} = -i$ $(S_x > 0)$ it gives two right-going chiral modes in the y direction. These modes are schematically shown in Fig. 3.9. Therefore it is natural to generate two Dirac cones in the junction by continuity. Thus these gapless states are protected by the mirror symmetry. If the mirror symmetry is not preserved the gapless interface states do not exist in general. Therefore our study reveals a new class of the interface states originating from the mirror symmetry.

As we have seen, the interface gapless states form a set of Dirac cones. The distance between the Dirac cones is proportional to the magnitude of the hybridization between the two TIs at the interface. When the hybridization becomes as strong as the bandwidth, the spacing between the interface Dirac cones is of the order of inverse of the lattice spacing. In that case the transport properties will be like graphene,

Table 3.1 Relations between the mirror Chern number $n_{\mathscr{M}}$ and the Chern numbers $n_{\pm i}$ in the mirror plane for each TIs

	$n_{\mathscr{M}}$	n_{+i}	n_{-i}
TI1	+1	+1	−1
TI2	−1	−1	+1

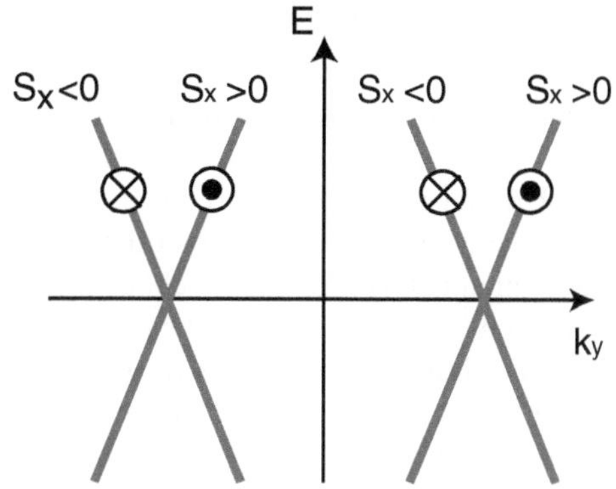

Fig. 3.9 Schematic of the dispersion of interface states in the mirror plane $k_x = 0$

having two Dirac cones at K and K′ points. We note that in graphene there are spin-degenerate Dirac cones, whereas in the present case the interface Dirac cones are not spin degenerate. From Fig. 3.9, when the wavenumber **k** goes around one of the Dirac point, the spin direction also rotates around the z axis (normal to the interface). In the similar way as in graphene, one can consider the valley degree of freedom as a pseudospin, and develop valleytronics [21–23] similar to graphene.

From the spin-resolved angle-resolved photoemission spectra, materials identified as TIs, such as $Bi_{1-x}Sb_x$ [24, 25], Bi_2Se_3[13], and Bi_2Te_3[14], have $n_{\mathcal{M}} = -1$; namely those chiralities are the same because the surface Dirac cone has spin texture with spins being clockwise in the k space. To realize the protected interface states in experiments discussed in this study, one needs to find a TI with $n_{\mathcal{M}} = +1$; i.e. the surface Dirac cone with negative velocity, and the spins on the upper cone is in the counterclockwise direction in the **k** space. It is an interesting issue to search for such TIs. The Dirac velocity v corresponds to the coefficient λ in the Rashba spin-splitting term $\lambda(\boldsymbol{\sigma} \times \mathbf{k})_z$ in the Hamiltonian, which is realized in surface [26] and in semiconductor heterostructure [27]. The Rashba coefficient λ originates from an integral of a sharply peaked function near the nuclei, which rapidly varies between positive and negative values [28, 29]. In recent studies [30, 31] the sign of the Rashba term depends on the material; the sign is plus in Au(111) surface, and is minus in Bi/Cu(111). Moreover in a recent experiment $Bi_4Se_{2.6}S_{0.4}$ has the opposite chirality [32]. Experimental investigations to find materials whose chirality is negative and theoretical studies of its origin would be promising.

Related to the issue discussed above, another class of topological phases called topological crystalline insulators has been proposed. Its surface states are protected by mirror symmetry is observed experimentally [33, 34]. Similar to our interface states, in those systems, there are an even number of Dirac surface states. While they belong to the weak topological insulator class, the gapless nature is preserved even without time-reversal symmetry, as long as the mirror symmetry does not broken.

3.3 Gapless Interface States in a Lattice Model

In the previous section we gave a general proof of the gapless interface states between two TIs. In this section, to show the interface states, we study the mirror Chern number and interface energy bands in a lattice model [3]. To this end, we use the Fu-Kane-Mele (FKM) tight-binding model on a diamond lattice with the spin-orbit interaction (SOI) [9]. The model exhibits Z_2 trivial and nontrivial phases. We find a relation between the mirror Chern number and the parameter values in the model. We then show the gapless states at the interface between two regions whose mirror Chern numbers are different from each other, in agreement with the results in the previous section.

3.3.1 Model

The FKM tight-binding Hamiltonian [9] is represented as

$$H = \sum_{\langle ij \rangle} t_{ij} c_i^\dagger c_j + \frac{8\lambda i}{a} \sum_{\langle\langle ij \rangle\rangle} c_i^\dagger \mathbf{s} \cdot (\mathbf{d}_{ij}^1 \times \mathbf{d}_{ij}^2) c_j. \tag{3.29}$$

c_i ($c_i^\dagger$) is the annihilation (generation) operator of the electron at i-th site. t_{ij}, λ are constants and a is the lattice spacing. The first term describes the nearest-neighbor hopping, which is independent of spin. The second term is the next-nearest-neighbor hopping term, which introduces the SOI in the system. $\mathbf{s}$ is the Pauli matrices for spins, and $\mathbf{d}_{ij}^{1,2}$ are vectors which constitute the next-nearest-neighbor bond from a site i to j.

The nearest-neighbor bond vectors (Fig. 3.10) are

$$\begin{cases} \boldsymbol{\tau}_1 = \frac{1}{4}(1, 1, 1), & \boldsymbol{\tau}_2 = \frac{1}{4}(-1, 1, -1), \\ \boldsymbol{\tau}_3 = \frac{1}{4}(-1, -1, 1), & \boldsymbol{\tau}_4 = \frac{1}{4}(1, -1, -1). \end{cases} \tag{3.30}$$

The next-nearest-neighbor bond vectors are

$$\begin{cases} \mathbf{d}_1 = \boldsymbol{\tau}_1 - \boldsymbol{\tau}_2 = \frac{1}{2}(1, 0, 1), & \mathbf{d}_2 = \boldsymbol{\tau}_1 - \boldsymbol{\tau}_3 = \frac{1}{2}(1, 1, 0), \\ \mathbf{d}_3 = \boldsymbol{\tau}_1 - \boldsymbol{\tau}_3 = \frac{1}{2}(0, 1, 1), & \mathbf{d}_4 = \boldsymbol{\tau}_2 - \boldsymbol{\tau}_3 = \frac{1}{2}(0, 1, -1), \\ \mathbf{d}_5 = \boldsymbol{\tau}_2 - \boldsymbol{\tau}_4 = \frac{1}{2}(-1, 1, 0), & \mathbf{d}_6 = \boldsymbol{\tau}_3 - \boldsymbol{\tau}_4 = \frac{1}{2}(-1, 0, 1). \end{cases} \tag{3.31}$$

Mirror symmetry is imposed with respect to the $(1\bar{1}0)$ plane by setting two of the four nearest-neighbor hoppings to be identical: $t_{\boldsymbol{\tau}_2} = t_{\boldsymbol{\tau}_4} = t_0$. For notational brevity, the other two hoppings are expressed as $t_{\boldsymbol{\tau}_1} = t_1$, $t_{\boldsymbol{\tau}_3} = t_2$.

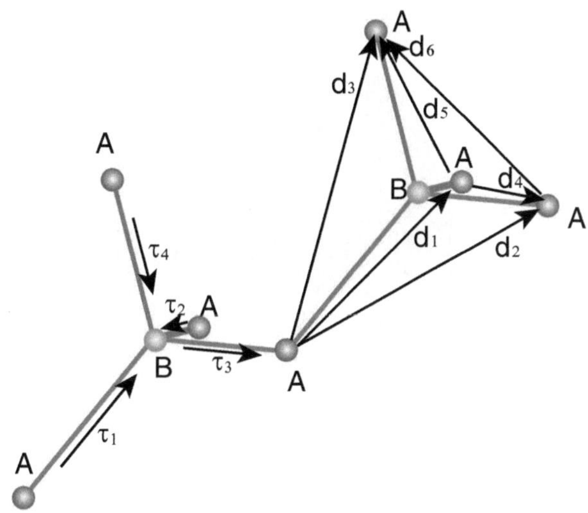

Fig. 3.10 Vectors for nearest-neighbor and next-nearest-neighbor hoppings in the FKM model on the diamond lattice

Under this condition, the Hamiltonian in the bulk is expressed as

$$H(\mathbf{k}) = \begin{pmatrix} 2\lambda s_i f_i(\mathbf{k}) & f_0(\mathbf{k}) \\ f_0^*(\mathbf{k}) & -2\lambda s_i f_i(\mathbf{k}) \end{pmatrix}, \tag{3.32}$$

where we use the summation convention over $i = 1, 2, 3$, and

$$\begin{cases} f_0(\mathbf{k}) = t_2 + t_0(\mathrm{e}^{-ik_4} + \mathrm{e}^{ik_6}) + t_1\mathrm{e}^{-ik_2}, \\ f_1(\mathbf{k}) = \sin k_1 - \sin k_2 + \sin k_5 - \sin k_6, \\ f_2(\mathbf{k}) = \sin k_2 - \sin k_3 - \sin k_4 + \sin k_5, \\ f_3(\mathbf{k}) = -\sin k_1 + \sin k_3 - \sin k_4 - \sin k_6, \end{cases} \tag{3.33}$$

$$k_j = \mathbf{d_j} \cdot \mathbf{k}. \tag{3.34}$$

The Brillouin zone (BZ) and dispersions for various sets of parameters are shown in Fig. 3.11. At $t_0 = t_2 = 1$ the model shows trivial and non-trivial Z_2 phases by tuning t_1. The model for $t_1 = 0.6$ in Fig. 3.11c is in the weak TI phase. On the other hand, for $t_1 = 1.4$ in Fig. 3.11d the model is classified as the strong TI phase [9]. The phase transition between them occurs through the bulk gapless states at $t_1 = 1$ $(= t_0 = t_2)$ (Fig. 3.11b).

3.3.2 Calculation of the Mirror Chern Number for the Model

Time-reversal invariant systems are classified by the Z_2 topological numbers. TIs belong to the non-trivial phases in the classification. When the system has a mirror

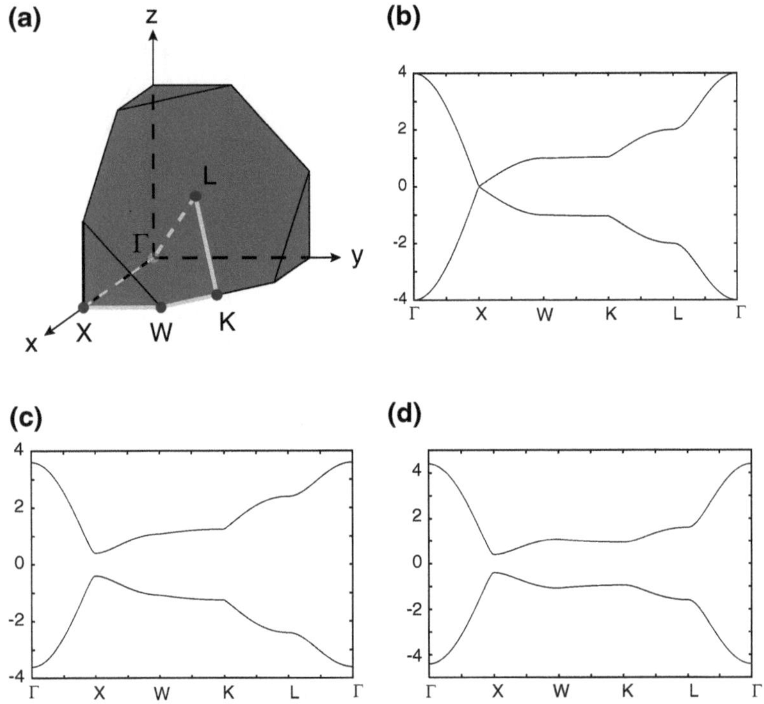

Fig. 3.11 **a** BZ for the diamond lattice structure in $k_x, k_y, k_z > 0$ and high symmetry points. (**b**, **c**, **d**) Bulk energy bands of the FKM model along the contour in (**a**). We set $t_2 = t_0 = 1.0$, and $\lambda = 0.125$, and **b** $t_1 = 1.0$, **c** $t_1 = 0.6$, **d** $t_1 = 1.4$

symmetry, the non-trivial phase is further classified by the mirror Chern number [18] which is related to the chirality of the surface states. The mirror operator is expressed as $\mathscr{M} = C_2 P$ where C_2 is the twofold rotation, and P is the parity. If the wavevector is restricted to be within the mirror plane, the model Hamiltonian (3.32) can be transformed into a block-diagonal form by the unitary operator $U_{\mathscr{M}}$, which diagnalizes $\mathscr{M}$ as $U_{\mathscr{M}} \mathscr{M} U^{\dagger}_{\mathscr{M}} = \text{diag}(+i, +i, -i, -i)$. As shown in the previous section, we characterize the topological phases in the FKM model using the mirror Chern number.

In the present case, the axis of the C_2 rotation is along $(1, -1, 0)$. The unitary matrix $U_{\mathscr{M}}$ which diagonalizes $\mathscr{M}$ is given by

$$U_{\mathscr{M}} = \begin{pmatrix} \hat{V} & 0 \\ 0 & \hat{V} \end{pmatrix}, \tag{3.35}$$

$$\hat{V} = \frac{1}{\sqrt{2}} \begin{pmatrix} -e^{i\frac{\pi}{4}} & e^{i\frac{\pi}{4}} \\ 1 & 1 \end{pmatrix}. \tag{3.36}$$

By $U_{\mathscr{M}}$, the Hamiltonian on the mirror plane $k_x = k_y = q$ is transformed as

$$U_{\mathscr{M}} H U_{\mathscr{M}}^{\dagger} = \begin{pmatrix} -2\sqrt{2}\lambda f_1 s_3 & f_0 \\ f_0^* & 2\sqrt{2}\lambda f_1 s_3 \end{pmatrix}, \tag{3.37}$$

where $f_0(\mathbf{k}) = t_2 + 2t_0 \exp\left(-i\frac{q-k_z}{2}\right) + t_1 \mathrm{e}^{-iq}$ and $f_1(\mathbf{k}) = \sin\left(\frac{q+k_z}{2}\right) - \sin q + \sin\left(\frac{q-k_z}{2}\right)$. Then this can be cast into the block-diagonalized Hamiltonian $H_{\mathscr{M}} = \mathrm{diag}(H_+, H_-)$, where

$$H_{\eta=\pm} = \begin{pmatrix} -2\sqrt{2}\eta\lambda f_1 & f_0 \\ f_0^* & 2\sqrt{2}\eta\lambda f_1 \end{pmatrix}. \tag{3.38}$$

By comparing the two Hamiltonians H_+ and H_-, the difference is only in the signs of the diagonal terms. Furthermore, these are transformed to each other by changing the sign of the SOI λ. The Chern numbers of the respective Hamiltonians $H_\pm$ have opposite signs from each other due to the time-reversal symmetry. Therefore, we conclude that the sign of λ corresponds to the sign of mirror Chern number $n_{\mathscr{M}}$.

To complement the above argument, we calculate directly the mirror Chern numbers, i.e. we compute the Chern numbers on the mirror plane. We use two methods for deriving the Chern number, which turn out to give identical results. One way (A) is to calculate a change of the Chern number across the phase transitions by tuning some parameters in H_η. This method enables us to calculate the Chern number analytically. The other way (B) to derive Chern number is to integrate the Berry curvature within the mirror plane in the BZ. Because the form of the Berry curvature becomes complicated, it is difficult, sometimes impossible, to calculate the Chern number analytically. Therefore we will numerically integrate the Berry phase to calculate the Chern numbers. In the following, we apply both methods (A) and (B) to the FKM model, and show that the results are the same. Details of these methods are given in Sects. 6.3 and 6.3.1.

3.3.2.1 (A) Calculation of the Difference of the Mirror Chern Numbers at the Phase Boundary in Parametric Space

To calculate the mirror Chern number for each region by the method (A), we need to search for points where the bulk gap closes by changing one external parameter. We introduce a 3D vector $\mathbf{k}$ consisting of the 2D wavevector k_1, k_2, and an external parameter denoted as k_3. k_3 can be any parameter that drives phase transition and in the present case k_3 will be set as one of the nearest-neighbor hoppings t_i. Suppose the bulk gap closes at $\mathbf{k} = \mathbf{k_0}$. Then, near the point $\mathbf{k} = \mathbf{k_0}$ the Hamiltonian is expressed as $H = s_i \alpha_{ij} \delta k_j$ in the lowest order in $\delta\mathbf{k} \equiv \mathbf{k} - \mathbf{k}_0$, if we restrict ourselves to two bands which are involved in the gap closing. Then by changing the parameters k_3

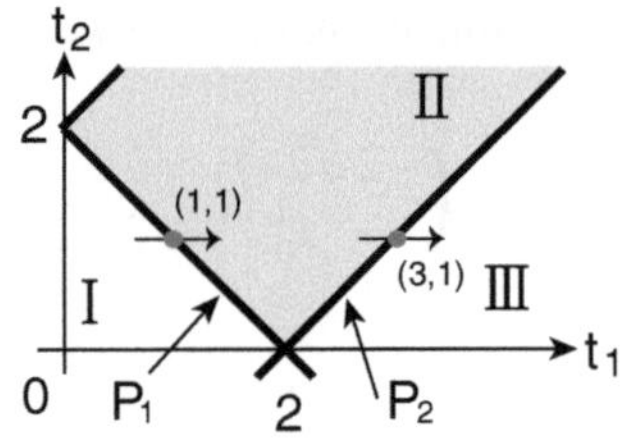

Fig. 3.12 Phase diagram in the (t_1, t_2) parametric space in units of t_0. When the parameters change across the boundaries, the mirror Chern number changes. In the argument given in the text, the mirror Chern numbers for Region *I* and *III* are zero, and $-\text{sign}(\lambda)$ for Region *II*

across the gap-closing point, we can show that the change of the Chern number δn has the form

$$\delta n = \text{sign}[\det(\alpha)]. \tag{3.39}$$

The method only gives the difference of the Chern numbers between neighboring areas in the parametric space. Therefore we have to know the Chern number at least in one region to determine the Chern number for other regions.[2] The mirror Chern number varies when the parameters traverse across the bulk gap closing on the mirror plane. This gap closing corresponds to a phase transition. Therefore we first need to identify the phase boundaries in the FKM model. In (t_0, t_1, t_2) parametric space, the gap closing occurs at

$$t_2 \pm t_1 \pm 2t_0 = 0. \tag{3.40}$$

From Eq. (3.40), the phase diagram (Fig. 3.12) is obtained for positive t_0, t_1 and t_2. Then we can derive the Chern numbers on the mirror plane by calculating their difference at the boundary. The region $t_2 + t_1 - 2t_0 < 0$ (Region I in Fig. 3.12) has the mirror Chern number $n_{\mathscr{M}} = 0$, because as the SOI is reduced to zero the region does not undergo any phase transition and remains gapped. In the limit of zero SOI, the Chern number $n_{\pm i}$ is zero and $n_{\mathscr{M}} = 0$. Therefore, even if the SOI is nonzero the mirror Chern number is zero in Region I.

For $t_2 - t_1 + 2t_0 < 0$ (in Region III in Fig. 3.12), the bulk is completely gapped, and the gap gets bigger with t_1 monotonically. On the other hand in Region II, there is a possibility that the mirror Chern number is not zero.[3] To obtain the mirror Chern

[2] For example, in QH systems the Chern number becomes zero when there is neither magnetization nor magnetic field. Starting from such a regime, one can determine the Chern number in other regimes in the parameter space.

[3] Actually the mirror Chern number is not zero, because Region II is the Z_2 non-trivial phase as shown in [9].

number we have to consider the two boundaries: $t_2-t_1+2t_0=0$ and $t_2+t_1-2t_0=0$. We call the boundary $t_2+t_1-2t_0=0$ and $t_2-t_1+2t_0=0$ P_1 and P_2 respectively.

On the phase boundary P_1, the bulk gap closes at the X point X^z $(0,0,2\pi)$ in the BZ. In the vicinity of X^z, we express the parameters as

$$\begin{cases} t_1 = t_0 + \delta t, \\ t_2 = t_0. \end{cases} \tag{3.41}$$

For $\delta t/t_0 \ll 1$, H_η is expressed in linear order in $\mathbf{p}=(p_1,p_2,p_3)=(\delta t,\delta q,\delta k_z)$ as

$$H_\eta = p_1 s_1 + p_3 s_2 + 4\sqrt{2}\eta\lambda p_2 s_3, \tag{3.42}$$

where

$$\begin{cases} \delta q = q, \\ \delta k_z = k_z - 2\pi. \end{cases} \tag{3.43}$$

Then we have

$$\alpha_\eta = \begin{pmatrix} 1 & 0 & 0 \\ 0 & 0 & 1 \\ 0 & 4\sqrt{2}\eta\lambda & 0 \end{pmatrix}. \tag{3.44}$$

From Eqs. (3.39) and (3.42), the change of the Chern numbers $\delta n_\eta|^{X^z}$, when δt changes across zero, is given as

$$\delta n_{\eta i}|^{X^z} = -\mathrm{sign}(\eta\lambda). \tag{3.45}$$

Because $n_{+i}=0$ in Region I, in Region II, we have $n_{+i}=-\mathrm{sign}(\lambda)$, $n_{-i}=\mathrm{sign}(\lambda)$ and $n_{\mathscr{M}}=-\mathrm{sign}(\lambda)$

On the phase boundary P_2, the gap closes at the L point (π,π,π). Around the L point, we express the parameters as

$$\begin{cases} t_1 = 3t_0 + \delta t, \\ t_2 = t_0, \\ q = \pi + \delta q, \\ k_z = \pi + \delta k_z. \end{cases} \tag{3.46}$$

The Hamiltonian is given in terms of $\mathbf{p}=(p_1,p_2,p_3)=(\delta t,\delta q,\delta k_z)$ as

$$H_\eta = -p_1 s_1 - (2p_2 - p_3)s_2 - 2\sqrt{2}\eta\lambda(p_2 - p_3)s_3. \tag{3.47}$$

Then α becomes

$$\alpha_\eta = \begin{pmatrix} -1 & 0 & 0 \\ 0 & -2 & 1 \\ 0 & -2\sqrt{2}\eta\lambda & 2\sqrt{2}\eta\lambda \end{pmatrix}, \tag{3.48}$$

and, the change of the Chern number across $\delta t = 0$ is

$$\delta n_{\eta i}|^{X^z} = \mathrm{sign}(\eta\lambda). \tag{3.49}$$

It agrees with the previous result from P_1 that the mirror Chern number is $-\mathrm{sign}(\lambda)$ in Region II. From these results, we have the mirror Chern number -1 ($+1$) in Region II for $\lambda > 0$ ($\lambda < 0$).

3.3.2.2 (B) Numerical Calculation of the Mirror Chern Numbers

Here we calculate the mirror Chern number, by (B) numerical integration of the Berry curvature within each mirror sector. The Hamiltonian labeled with the mirror eigenvalue η (Eq. 3.38) can be expressed as

$$H_\eta = \beta_{\eta j} s_j, \tag{3.50}$$

$$\begin{cases} \beta_{\eta 1} = 2t_0 \cos\left(\frac{q-k_z}{2}\right) + t_1 \cos q + t_2, \\ \beta_{\eta 2} = 2t_0 \sin\left(\frac{q-k_z}{2}\right) + t_1 \sin q, \\ \beta_{\eta 3} = -2\eta\sqrt{2}\lambda f_1. \end{cases} \tag{3.51}$$

From the above, the Chern number n_η for the subspace with mirror eigenvalues η is given as

$$n_{\eta i} \equiv \frac{1}{4\pi} \int_{BZ}\int \mathrm{d}q\mathrm{d}k_z B_\eta(q, k_z), \tag{3.52}$$

$$B_\eta(q, k_z) = \hat{\boldsymbol{\beta}}_\eta \cdot \frac{\partial \hat{\boldsymbol{\beta}}_\eta}{\partial q} \times \frac{\partial \hat{\boldsymbol{\beta}}_\eta}{\partial k_z}, \tag{3.53}$$

where $B_\eta(q, k_z)$ is the Berry curvature, and $\hat{\boldsymbol{\beta}}_\eta = \boldsymbol{\beta}_\eta/|\boldsymbol{\beta}_\eta|$. The distribution of $B_{+1}(q, k_z)$ is calculated in Fig. 3.13. By the time-reversal symmetry we have $B_{+1}(q, k_z) = -B_{-1}(-q, -k_z)$. Integrating $B_{+1}(q, k_z)$ over the BZ, we can calculate n_+ for the individual values of (t_1, t_2) (Fig. 3.14). The resulting mirror Chern number is approximately -1.0 in Region II for $\lambda = 0.125 (>0)$ and agrees with the previous result.

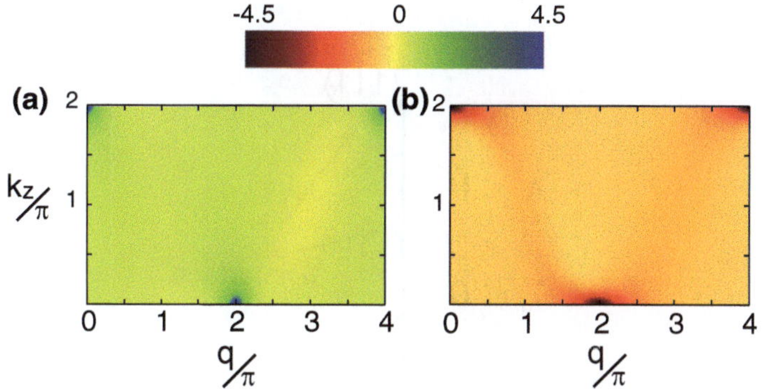

Fig. 3.13 Distribution of the Berry curvature, $B_+(q, k_z)$ in the BZ. The parameters are $t_2 = t_0 = 1.0$, and $\lambda = 0.125$, $t_1 = 1.4$, and **a** $t_1 = 0.6$ (Region I), **b** $t_1 = 1.4$ (Region II). The distribution of the Berry curvature concentrates more on X^z $(0, 0, 2\pi)$ by getting closer to the phase boundary ($t_1 = 1.0$), and the Berry curvature discretely jumps at the transition

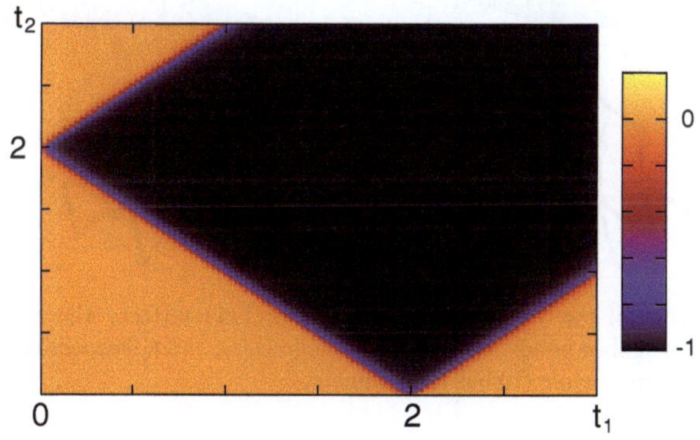

Fig. 3.14 Distribution of the Chern number for the $\eta = +1$ sector in (t_1, t_2) parametric space by numerical calculation for $\lambda = 0.125$. In Region I and III (in Fig. 3.12), the Chern number $n_+ \sim 0.0$, while in II $n_+ \sim -1.0$

3.3.3 Gapless Interface States

In the previous section and Sect. 3.2, we found that the sign of λ in the FKM model corresponds to the sign of the mirror Chern number and the chirality of the surface states. Therefore, we expect the existence of the gapless interface states on an interface between two regions having opposite signs of λ. In this section we numerically calculate the interface states for this junction. In order to avoid appearance of the surface states in the junction system, we impose periodic boundary condition along the surface normal ((111) direction) as shown in Fig. 3.15, and the wavenumber along (111) is set to be zero to focus on the 2D interface plane.

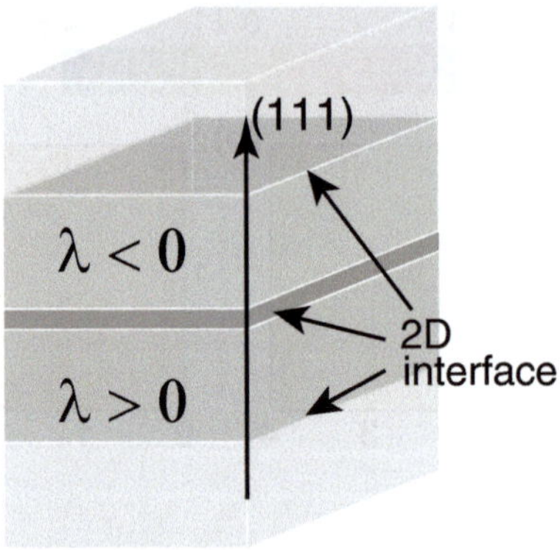

Fig. 3.15 Schematic of the junction system between two TIs with opposite signs of the SOI (λ) in the FKM model. We impose boundary condition along the (111) direction

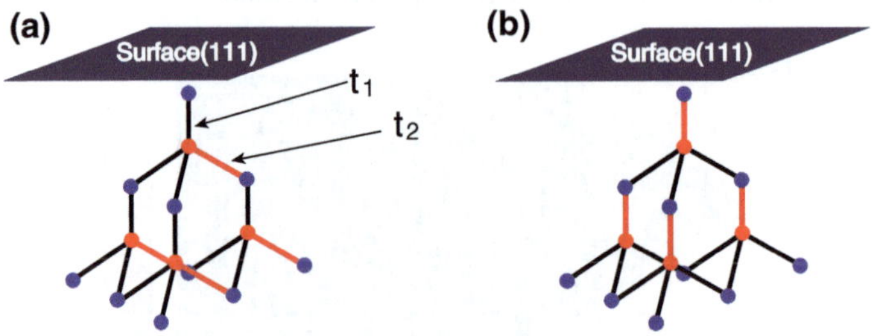

Fig. 3.16 Schematics of the diamond lattice with the (111) surface. The *black* and *red* bonds denote the nearest-neighbor bonds with hopping integrals, t_1, and t_2, respectively. In (**a**), there are no rotational symmetries, while (**b**) has C_3 symmetry

Before calculating the interface states, we note rotational symmetries in the crystalline system considered. When $t_0 = t_1 \neq t_2$ (Fig. 3.16a), there is no rotational symmetry. On the other hand, when $t_0 = t_2$ (Fig. 3.16b), there is C_3 symmetry with respect to the (111) axis.

The calculated dispersions are shown in Fig. 3.17. When the system does not have the C_3 symmetry, there are two Dirac cones in the dispersion as shown in Fig. 3.17a for $(t_1, t_2) = (1, 1.4)$. On the other hand for $(t_1, t_2) = (1.4, 1)$, there are six Dirac cones (Fig. 3.17b) due to the C_3 symmetry. Two Dirac points are located on the mirror plane in the BZ in both of the results. It agrees with the argument with the mirror Chern number, because the difference of $n_{\mathscr{M}}$ between regions in the system is $+1-(-1) = +2$. The entire dispersion reflects the spatial symmetry, i.e. whether or not the system has C_3 symmetry. The results are in accordance with the study in the previous section.

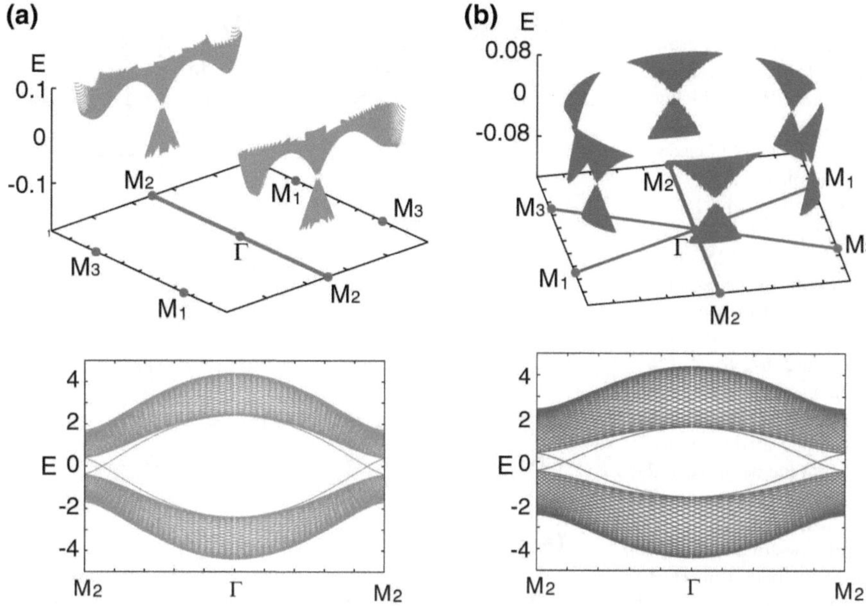

Fig. 3.17 Energy bands of the gapless interface states in the interface BZ (Fig. 3.16) for $|\lambda| = 0.125t_0$. The energy E and the parameters $t_{1,2}$ are represented in units of t_0. Only the states near the Fermi energy ($E = 0$) are shown. Here the bulk bands do not appear because they are gapped and are away from the Fermi energy $E = 0$. The mirror plane corresponds to the M_2-Γ-M_2 line. The dispersions in the mirror plane M_2-Γ-M_2 line are shown in the *lower panels*. **a** There are two Dirac cones when the system does not have rotational symmetry. The parameters are $t_1 = 1.0$ and $t_2 = 1.4$. **b** There are six Dirac cones, when the system has C_3 symmetry. The parameters are $t_2 = 1.0, t_1 = 1.4$

3.4 Interfacial Fermi Loops and Interfacial Symmetries

3.4.1 Interfacial Fermi Loops in the FKM Model

Previously, we have shown the metallic interface states protected by mirror symmetry by using the FKM model. Next, we show that when $\lambda_\alpha = -\lambda_\beta$, a novel dispersion of interface states appears near the Fermi energy ($E = 0$), as shown in Fig. 3.18, which is in contrast with Fig. 3.17. The parameters of the SOI are $\lambda_\alpha = -\lambda_\beta = 0.125t_0$ for all the results, and (t_1, t_2) is given as $(1.4, 1)t_0$ in Fig. 3.18a-1, and $(1, 0.6)t_0$ in Fig. 3.18b-1. Figure 3.18a-1 belongs to the TI phase, and Fig. 3.18b-1 is in the non-TI phase. In all the cases, the gap closes at $E = 0$, with its degeneracy forming loops in the interface BZ. The dispersion is linear in the direction perpendicular to the loop. Remarkably, this loop is not on a high-symmetry line. This kind of degeneracy along a loop is unexpected in general lattice models, because there is no symmetry

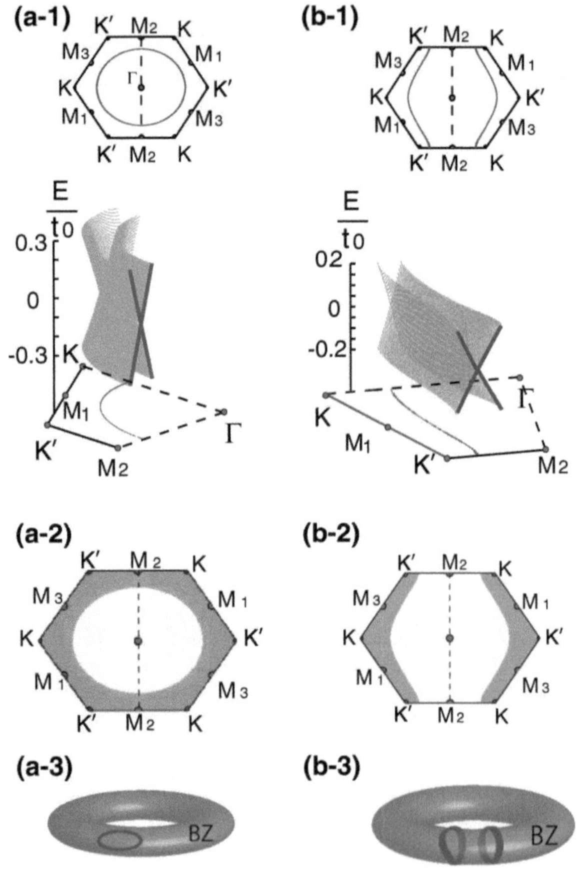

Fig. 3.18 **a-1, b-1** Dispersion and the Fermi surface for the interface states with IPHS. Thickness of each region α, β is 24 unit cells along the (111) axis. The parameters are (a) $(t_1, t_2) = (1.4, 1)t_0$, (b)$(t_1, t_2) = (1, 0.6)t_0$, and $\lambda_\alpha = -\lambda_\beta = 0.125t_0$ for (a) and (b). In the BZ in each panel, the Fermi surfaces at $E = 0$ are shown as *blue curves*. **a-2, b-2** Sign of the Pfaffian sgn(pf($\tilde{U}H$)) in the interface BZ. The *shaded regions* and *white regions* represent the negative and positive signs, respectively. Schematic illustrations of the Fermi loops of a closed orbit (**a-3**) and open orbits (**b-3**) are shown on the torus of the interface BZ

which protects degeneracy between the valence and conduction bands along a loop. Therefore, new theoretical description is required [3].

For the sake of explaining this peculiar dispersion, we consider a transformation of the FKM Hamiltonian as

$$UH_{\lambda,\mathbf{k}}U^{-1} = -H^t_{-\lambda,-\mathbf{k}}, \tag{3.54}$$

where $U = i\tau_z s_y$. The Hamiltonian also has an inversion symmetry: $PH_{\lambda,\mathbf{k}}P^{-1} = H_{\lambda,-\mathbf{k}}$, $P = \tau_x$. An operator $UK = i\tau_z s_y K$ corresponds to the following sublattice-dependent PH transformation: $\tilde{c}^\dagger_{\alpha i\pm} = \pm c_{\alpha i\mp}$, $\tilde{c}^\dagger_{\beta i\pm} = \mp c_{\beta i\mp}$, which leaves the spin operator invariant: $c^\dagger_i \mathbf{s} c_i = \tilde{c}^\dagger_i \mathbf{s}\tilde{c}_i$. The respective regions α and β are not invariant under the PH transformation, but they are transformed to each other by this PH transformation because $\lambda_\alpha = -\lambda_\beta$. Therefore, by making the α-β interface, the PHS is restored as shown in the following.

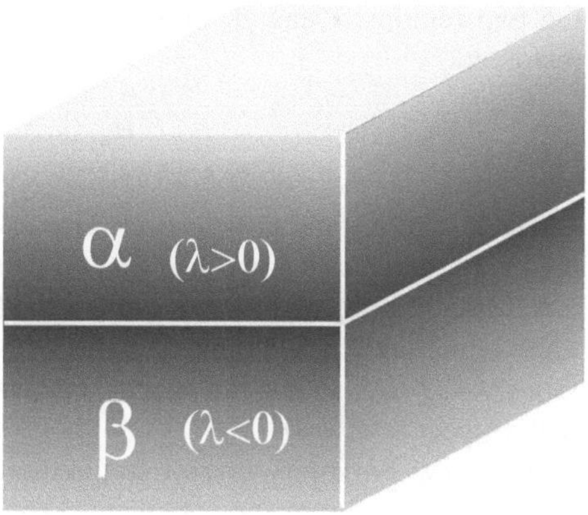

Fig. 3.19 Schematic of the junction system labeled with α and β. For the FKM model, these labels correspond to signs of the SOI

The existence of the anomalous interface states forming a Fermi loop is explained as follows. First we note that for a Hamiltonian $H^{2D}_{\lambda,\mathbf{k}_\parallel}$ describing a 2D slab with finite thickness, we have $UH^{2D}_{\lambda,\mathbf{k}_\parallel}U^{-1} = -(H^{2D}_{-\lambda,-\mathbf{k}_\parallel})^t$ from Eq. (3.54), where $\mathbf{k}_\parallel$ is a 2D wavevector parallel to the interface. Then the junction of two regions α, β (Fig. 3.19a), with their SOI parameters given by $\lambda_\alpha = -\lambda_\beta = \lambda$ is described by the Hamiltonian

$$H_{\mathbf{k}_\parallel} = \begin{pmatrix} H^{2D}_{\lambda,\mathbf{k}_\parallel} & V_{\mathbf{k}_\parallel} \\ V^\dagger_{\mathbf{k}_\parallel} & H^{2D}_{-\lambda,\mathbf{k}_\parallel} \end{pmatrix}, \tag{3.55}$$

where V is the hybridization between the regions α and β. We have the following relation,

$$UH_{\mathbf{k}_\parallel}U^{-1} = \begin{pmatrix} -(H^{2D}_{-\lambda,-\mathbf{k}_\parallel})^t & UV_{\mathbf{k}_\parallel}U^{-1} \\ UV^\dagger_{\mathbf{k}_\parallel}U^{-1} & -(H^{2D}_{\lambda,-\mathbf{k}_\parallel})^t \end{pmatrix}. \tag{3.56}$$

Therefore, the SOI parameter λ changes sign, and the regions α and β are exchanged. When the two regions α, β have the same thickness, we can exchange the regions again by the space inversion. The space inversion is given by $PP'\Sigma_x$, where P' is an operator for inversion of the stacking direction of layers within each region and Σ_x is the Pauli matrix acting to the region degrees of freedom, α and β. When each region α and β have N AB pairs along the stacking direction, P' is given as $P'_{ij} = \delta_{i,N+1-j}\tau_0 s_0$, where i and j denote the indices for the sublattice pair from the top, and τ_0 and s_0 are the identity operators for the sublattice and spin spaces. Therefore when the hybridization V satisfies

$$PP'UV_{\mathbf{k}_\parallel}(PP'U)^{-1} = -V^t_{\mathbf{k}_\parallel}, \tag{3.57}$$

the two regions α and β are exchanged by the space inversion, and the PHS is restored:

$$\tilde{U} H_{\mathbf{k}_{\parallel}} \tilde{U}^{-1} = -H^t_{\mathbf{k}_{\parallel}}, \tag{3.58}$$

where $\tilde{U} = Q\Sigma_x$ and $Q = PP'U$. Thus, although the Hamiltonian in each region is not invariant under a combined transformation of U and the inversion, by making a junction consisting of the two regions α, β, the system restores a symmetry under a PH transformation U and the spatial inversion, which we call *the interfacial particle-hole symmetry* (*IPHS*). Thus the energy eigenvalues of the whole system are symmetric with respect to $E = 0$.

This new symmetry gives rise to emergence of the Fermi loop in the interface states, as we show in the following. From Eq. (3.58), $\tilde{U} H_{\mathbf{k}_{\parallel}}$ is a skew symmetric matrix,

$$[\tilde{U} H_{\mathbf{k}_{\parallel}}]^t = H^t_{\mathbf{k}_{\parallel}} \tilde{U}^t = -\tilde{U} H_{\mathbf{k}_{\parallel}} \tilde{U}^{-1} \tilde{U}^t = -\tilde{U} H_{\mathbf{k}_{\parallel}}, \tag{3.59}$$

and furthermore its Pfaffian is real:

$$\begin{aligned} \mathrm{pf}[\tilde{U} H_{\mathbf{k}_{\parallel}}] &= \mathrm{pf}[-H^t_{\mathbf{k}_{\parallel}} \tilde{U}^t] = \det(\tilde{U}) \mathrm{pf}[\tilde{U}^{-1} H^t_{\mathbf{k}_{\parallel}}] \\ &= \mathrm{pf}[\tilde{U} H_{\mathbf{k}_{\parallel}}]^*. \end{aligned} \tag{3.60}$$

To derive these relations, we used the fact that the matrix $\tilde{U}$ is unitary and symmetric, and the fact that the size of the matrix is an integer multiple of four. The reality condition of $\mathrm{pf}[\tilde{U} H_{\mathbf{k}_{\parallel}}]$ gives a strong constraint for the Fermi surface. Suppose the sign of $\mathrm{pf}(\tilde{U} H_{\mathbf{k}_{\parallel}})$ changes at some $\mathbf{k}_{\parallel}$; it then means $\mathrm{pf}(\tilde{U} H_{\mathbf{k}_{\parallel}}) = 0$, i.e. $\det\tilde{U}$ $\det H_{\mathbf{k}_{\parallel}} = 0$ and $\det H_{\mathbf{k}_{\parallel}} = 0$. Therefore, the Hamiltonian $H_{\mathbf{k}_{\parallel}}$ has zero eigenvalues and the gap closes at this wavevector $\mathbf{k}_{\parallel}$, because of the PH symmetry. Because the wavevector $\mathbf{k}_{\parallel}$ where the sign of $\mathrm{pf}(\tilde{U} H_{\mathbf{k}_{\parallel}})$ changes forms a loop in the 2D BZ, this loop is nothing but the Fermi loop where the gap is closed. We note that we have set the thickness of the two regions α and β to be the same. This can be relaxed as long as the slab thickness is much larger than the decay length of the interface states.

To confirm that the Fermi loop shown in Fig. 3.18 is due to this scenario, we calculate the Pfaffian $\mathrm{pf}(\tilde{U} H)$ in this model by using PFAPACK [35]. The results are shown in Fig. 3.18a-2 and 3.18b-2 by using the same parameters for Fig. 3.18a-1 and 3.18b-1. Here the shaded and white regions represent $\mathrm{sgn}(\mathrm{pf}(\tilde{U} H_{\mathbf{k}_{\parallel}})) = -1$, and $\mathrm{sgn}(\mathrm{pf}(\tilde{U} H_{\mathbf{k}_{\parallel}})) = 1$, respectively. Hence the wavevectors $\mathbf{k}$ for the sign change of $\mathrm{pf}(\tilde{U} H_{\mathbf{k}_{\parallel}})$ agree with the interface Fermi loops shown in Fig. 3.18. In addition, the Fermi loops in (a-2) form closed orbits, while those in (b-2) do open orbits. Due to the reality of $\mathrm{pf}(\tilde{U} H)$, the Fermi loops are classified into two types: closed orbits and open orbits (Fig. 3.18a-3 and b-3). Because of the nontrivial winding number for the open orbits, it is not possible for the open-orbit Fermi loops to disappear without transition to closed loops. We note that these behaviors are similar to the Fermi surface of tight-binding models on the diamond lattice without the SOI [36].

To realize gapless topological interface states due to the mirror Chern number experimentally, two TIs with opposite chiralities are necessary and mirror symmetry normal to the interface is required in common. In a recent experiment, a natural superlattice, $Bi_4Se_{2.6}S_{0.4}$ is reported to have a surface Dirac cone with the chirality opposite from the conventional one [32], although it is not a TI but a semimetal. Furthermore, it might be interesting to search for interface Fermi loops between two insulators from this scenario.

3.4.2 Interfacial Fermi Loops from Interfacial Symmetries

Our theory on topological Fermi loops can be generalized.[4] We construct a system with two regions α, β (Fig. 3.19) forming 2D slabs. The bulk Hamiltonians in the respective regions are assumed to be related to each other by an operator U:

$$H_{\alpha(\beta),\mathbf{k}} = \varepsilon_C U H^t_{\beta(\alpha),-\mathbf{k}} U^{-1}, \tag{3.61}$$

where $\varepsilon_C = \pm 1$. We can classify the operator U as PHS-like for $\varepsilon_C = -1$ or TRS-like for $\varepsilon_C = 1$. As is similar to the FKM model, if the hybridization V between the two regions satisfies the condition, $\tilde{P} U V_{\mathbf{k}_\parallel} (\tilde{P} U)^{-1} = \varepsilon_C V^t_{\mathbf{k}_\parallel}$, where $\tilde{P}$ inverts the order of stacking along the interface normal, the Hamiltonian satisfies a relation

$$\tilde{U} H_{\mathbf{k}_\parallel} \tilde{U}^{-1} = \varepsilon_C H^t_{\mathbf{k}_\parallel}, \quad \tilde{U} \equiv \tilde{P} U \Sigma_x. \tag{3.62}$$

Thus, by making a junction between the regions α and β, the system restores a symmetry under a combined transformation of U and inversion. We call this symmetry IPHS for $\varepsilon_C = -1$ as mentioned previously, and *interfacial time-reversal symmetry (ITRS)* for $\varepsilon_C = 1$. These new symmetries are classified by $\tilde{U} = \eta_U \tilde{U}^t$, $\eta_U = \pm 1$. Because $[\tilde{U} H_{\mathbf{k}_\parallel}]^t = H^t_{\mathbf{k}} \tilde{U}^t = \varepsilon_C \eta_U \tilde{U} H_{\mathbf{k}_\parallel}$, $\tilde{U} H_{\mathbf{k}}$ is a skew symmetric matrix only for $\varepsilon_C \eta_U = -1$, i.e.

$$(\varepsilon_C, \eta_U) = (\mp 1, \pm 1). \tag{3.63}$$

Then, the Pfaffian $\mathrm{pf}(\tilde{U} H_{\mathbf{k}_\parallel})$ is defined and is real as before.

The reality of the Pfaffian $\mathrm{pf}(\tilde{U} H_{\mathbf{k}_\parallel})$ is reflected in the dispersions of interface states in a different way between the IPHS and the ITRS. For junction systems with the IPHS ($(\varepsilon_C, \eta_U) = (-1, 1)$) between two insulating regions, the gap between the valence and conduction bands closes along a Fermi loop, as we saw in the FKM model. It is usually unlikely for the gap to close at an interface between two insulators,

[4] Details of calculations in this section are given in Sects. 7.1 and 7.2.

and it is even more unlikely to close the gap along a loop. Nevertheless, for the present cases with the IPHS, if the gap of the interface states is closed somewhere, the gap should close not at an isolated point but along a loop due to the reality of the Pfaffian. In the example of the FKM model, there is a reason for the gap to close; namely the difference of the mirror Chern number between the two regions results in the presence of gap-closing points along the mirror plane. Together with the reality of the Pfaffian, the gap closes along a loop.

In fact the appearance of Fermi loops in the IPHS class is not limited to cases with topological interfacial states (from the mirror Chern number), but can be found in general junction systems with IPHS, as long as the hybridization at the interface is stronger than the gap. To show this, we consider a junction system with weak hybridization $V_{\mathbf{k}_\parallel}$ between two regions α and β. We diagonalize the Hamiltonian for the α-region, $H_{\alpha,\mathbf{k}_\parallel}$, by a unitary matrix $W_{\mathbf{k}_\parallel}$ as

$$W_{\mathbf{k}_\parallel}^\dagger H_{\alpha,\mathbf{k}_\parallel} W_{\mathbf{k}_\parallel} = \mathrm{diag}(E_{1\mathbf{k}_\parallel}, E_{2\mathbf{k}_\parallel}, \ldots, E_{\bar{N}\mathbf{k}_\parallel}), \tag{3.64}$$

where we assume $E_{i\mathbf{k}_\parallel} < E_{i+1\mathbf{k}_\parallel}$ for $i = 1, 2, \ldots, \bar{N} - 1$, and $\bar{N}$ is the size of the matrix. The two regions are band insulators, and the Fermi energy E_F (=0) is assumed to be between $E_{m\mathbf{k}_\parallel}$ and $E_{m+1\mathbf{k}_\parallel}$. By the interface transformation, the Hamiltonian for the β-region $H_{\beta,\mathbf{k}_\parallel}$ is diagonalized by $W'_{\mathbf{k}_\parallel} \equiv U^* W^*_{-\mathbf{k}_\parallel}$ as

$$W'^\dagger_{\mathbf{k}_\parallel} H_{\beta,\mathbf{k}_\parallel} W'_{\mathbf{k}_\parallel} = \mathrm{diag}(-E_{1,-\mathbf{k}_\parallel}, -E_{2,-\mathbf{k}_\parallel}, \ldots, -E_{\bar{N},-\mathbf{k}_\parallel}). \tag{3.65}$$

By assuming $|E_{m\mathbf{k}_\parallel}|, |E_{m+1\mathbf{k}_\parallel}| \ll |E_{i\mathbf{k}_\parallel}|$ for $i \neq m, m+1$, leading-order terms of the Pfaffian expanded in terms of V' are given as

$$\mathrm{pf}(\tilde{U} H_{\mathbf{k}_\parallel}) \sim Z_{\bar{N}} \left[1 + \frac{|V'_{m,m+1,-\mathbf{k}_\parallel}|^2}{E_{m,-\mathbf{k}_\parallel} E_{m+1,-\mathbf{k}_\parallel}} \right], \tag{3.66}$$

where $Z_{\bar{N}} = \prod_{i=1}^{\bar{N}} E_{i,-\mathbf{k}_\parallel}$. Since each region is assumed to be gapped for the entire BZ, the sign of $Z_{\bar{N}}$ is fixed. Thus, the condition for the Fermi loop is given as

$$E_{m\mathbf{k}_\parallel} E_{m+1\mathbf{k}_\parallel} = -|V'_{m,m+1,\mathbf{k}_\parallel}|^2. \tag{3.67}$$

It means that the Fermi loop appears when the hybridization becomes comparable to the gap for the IPHS. For example, when the chemical potential to be the middle of the gap ($-E_{m+1,\mathbf{k}_\parallel} = E_{m,\mathbf{k}_\parallel}$), the Fermi loop appears at $E_{m+1,\mathbf{k}_\parallel} = |V'_{m,m+1,\mathbf{k}_\parallel}|$.

On the other hand, for the ITRS class $(\varepsilon_C, \eta_U) = (1, -1)$, it results in double degeneracy for every eigenstate, because the eigenequation becomes a perfect square: $\mathrm{Det}(H - E) = \mathrm{Det}(\tilde{U}(H - E)) = (\mathrm{pf}(\tilde{U}(H - E)))^2$. Here we used the fact that $\tilde{U}$ is skew symmetric for the ITRS class. Therefore, although within each region the TRS is broken, the ITRS gives rise to a "Kramers-like" degeneracy for every state.

3.4.3 Interfacial Fermi Loops in the π-junction Interface

In general, it might be difficult to realize interfaces with the IPHS. On the other hand, it is easier for superconductors, because their energy bands have the PHS by themselves. Here we show another example of the Fermi loop at the π-junction interface between two 2D Rashba systems. We consider 2D bilayer systems with the Rashba SOI on the square lattice [37, 38], and assume that s-wave superconductivity is induced by a proximity effect. In the case with the monolayer Rashba system, the Bogoliubov-de Gennes Hamiltonian is written as

$$H_\Delta(\mathbf{k}) = \begin{pmatrix} h_0(\mathbf{k}) & is_y\Delta \\ -is_y\Delta & -h_0^t(-\mathbf{k}) \end{pmatrix}, \tag{3.68}$$

$$h_0(\mathbf{k}) = \xi(\mathbf{k}) + \alpha(s_x \sin k_y - s_y \sin k_x), \tag{3.69}$$

where Δ is the superconducting order parameter and is assumed to be a real constant, $\xi(\mathbf{k})$ $(= \xi(-\mathbf{k}))$ is the kinetic energy from the chemical potential μ, and α is the strength of the SOI. The eigenvalues are

$$E_\Delta = \pm\sqrt{\left(\xi(\mathbf{k}) \pm \alpha\sqrt{\sin^2 k_x + \sin^2 k_y}\right)^2 + \Delta^2}, \tag{3.70}$$

and the system is gapped for the entire BZ. Because the Hamiltonian satisfies the equations

$$\tau_y H_\Delta(\mathbf{k})\tau_y = -H^t_{-\Delta}(-\mathbf{k}),\ \tau_z s_z H_\Delta(\mathbf{k})\tau_z s_z = H_\Delta(-\mathbf{k}), \tag{3.71}$$

we can see that the Hamiltonians $H_{\pm\Delta}(\mathbf{k})$ are related by the PHS with $(\varepsilon_C, \eta_U) = (-1, 1)$. Therefore, a π-junction interface between two 2D Rashba systems with $+\Delta$ and $-\Delta$ has the IPHS, and its Hamiltonian is given by

$$H(\mathbf{k}) = \begin{pmatrix} H_\Delta(\mathbf{k}) & t\tau_z \\ t\tau_z & H_{-\Delta}(\mathbf{k}) \end{pmatrix}, \tag{3.72}$$

where t is the hybridization across the junction, and τ_z is the Pauli matrix for Nambu space. At the band edge, we approximate the kinetic energy as $\xi(\mathbf{k}) = \frac{k^2}{2m_e} - \mu$, where m_e is the electron mass, and $\sin k_i \sim k_i$. We, then, obtain the energy eigenvalues

$$E_\Delta = \pm t \pm \sqrt{\left(\frac{k^2}{2m_e} - \mu \pm \alpha k\right)^2 + \Delta^2}. \tag{3.73}$$

Therefore, the Fermi loops appear when the hybridization is larger than the gap ($|t| > |\Delta|$). It can be satisfied even if the junction is weak, because the hybridization t can be maximally $\sim 10^{-1}$ eV, and the gap Δ is typically $\Delta \sim 10^{-4}$ eV.

References

1. M.Z. Hasan, C.L. Kane, Rev. Mod. Phys. **82**, 3045 (2010)
2. R. Takahashi, S. Murakami, Phys. Rev. Lett. **107**, 166805 (2011)
3. R. Takahashi, S. Murakami, Phys. Rev. Lett. **113**, 256406 (2014)
4. K.S. Novoselov et al., Science **306**, 666 (2004)
5. C.L. Kane, E.J. Mele, Phys. Rev. Lett. **95**, 226801 (2005)
6. C.L. Kane, E.J. Mele, Phys. Rev. Lett. **95**, 146802 (2005)
7. B.A. Bernevig, T.L. Hughes, S.-C. Zhang, Science **314**, 1757 (2006)
8. J.E. Moore, L. Balents, Phys. Rev. B **75**, 121306(R) (2007)
9. L. Fu, C.L. Kane, E.J. Mele, Phys. Rev. Lett. **98**, 106803 (2007)
10. M.I. Katsnelson et al., Nat. Phys. **2**, 620–625 (2006)
11. V.V. Cheianov et al., Science **315**, 1252 (2007)
12. Y. Xia et al., Nat. Phys. **5**, 398 (2009)
13. D. Hsieh et al., Nature (London) **460**, 1101 (2009)
14. Y.L. Chen et al., Science **325**, 178 (2009)
15. V.G. Veselago, Sov. Phys. Usp. **10**, 509 (1968)
16. J.B. Pendry, Phys. Rev. Lett. **85**, 3966 (2000)
17. L. Fu, Phys. Rev. Lett. **103**, 266801 (2009)
18. J.C.Y. Teo, L. Fu, C.L. Kane, Phys. Rev. B **78**, 045426 (2008)
19. D.J. Thouless, M. Kohmoto, M.P. Nightingale, M. den Nijs, Phys. Rev. Lett. **49**, 405 (1982)
20. M. Kohmoto, Ann. Phys. (N.Y.) **160**, 343 (1985)
21. O. Gunawan et al., Phys. Rev. Lett. **97**, 186404 (2006)
22. D. Xiao, W. Yao, Q. Niu, Phys. Rev. Lett. **99**, 236809 (2007)
23. A. Rycerz et al., Nat. Phys. **3**, 172 (2007)
24. D. Hsieh et al., Nature (London) **452**, 970 (2008)
25. A. Nishide et al., Phys. Rev. B **81**, 041309(R) (2010)
26. S. LaShell, B.A. McDougall, E. Jensen, Phys. Rev. Lett. **77**, 3419 (1996)
27. J. Nitta, T. Akazaki, H. Takayanagi, T. Enoki, Phys. Rev. Lett. **78**, 1335 (1997)
28. G. Bihlmayer et al., Surf. Sci. **600**, 3888 (2006)
29. M. Nagano, A. Kodama, T. Shishidou, T. Oguchi, J. Phys.: Condens. Matter **21**, 064239 (2009)
30. J. Henk, M. Hoesch, J. Osterwalder, A. Ernst, P. Bruno, J. Phys. Condens. Matter **16**, 7581 (2004)
31. H. Bentmann et al., Phys. Rev. B **84**, 115426 (2011)
32. T. Valla et al., Phys. Rev. B **86**, 241101(R) (2012)
33. T.H. Hsieh et al., Nat. Commun. **3**, 982 (2012)
34. Y. Tanaka et al., Nat. Phys. **800**, 982 (2012)
35. M. Wimmer, ACM Trans. Math. Softw. **38**, 30 (2012)
36. R. Takahashi, S. Murakami, Phys. Rev. B **88**, 235303 (2013)
37. E. Bernardes, J. Schliemann, M. Lee, J.C. Egues, D. Loss, Phys. Rev. Lett. **99**, 076603 (2007)
38. M. Akabori et al., J. Appl. Phys. **112**, 113711 (2012)

Chapter 4
Weyl Semimetal in a Thin Topological Insulator

4.1 Gapless States Protected by Mirror Symmetry in a Topological Insulator

In the low-energy model, the surface sates of topological insulators (TIs) are expressed by the massless Dirac equation,

$$H_0 = v(\sigma_x k_y - \sigma_y k_x), \tag{4.1}$$

where v is the Dirac velocity. When a ferromagnetic film with magnetization $\mathbf{m}$ is attached to a surface of a three-dimensional (3D) TI, the surface state is gapped because the time-reversal symmetry is broken. The Hamiltonian H and eigenvalue E are expressed as

$$H = H_0 + H_m = \sigma_x(vk_y - m_x) - \sigma_y(vk_x + m_y) - m_z\sigma_z, \tag{4.2}$$

$$E(\mathbf{k}) = \pm\sqrt{(k_y - m_x)^2 + (k_x + m_y) + m_z^2}, \tag{4.3}$$

where $H_m = -\mathbf{m} \cdot \boldsymbol{\sigma}$, and$-\mathbf{m} \cdot \boldsymbol{\sigma}$ is the Zeeman field. From Eq. (4.3), we can see that the surface Dirac model is not gapped by the in-plane Zeeman field ($m_z = 0$), although the time-reversal symmetry is broken. This is due to the mirror symmetry. Namely, when the Zeeman field is in-plane, the mirror symmetry with respect to the plane perpendicular to the field is preserved. The in-plane Zeeman field causes energy shifts of the states on the mirror plane (Fig. 4.1). Within the mirror plane, the crossing between the two branches is not avoided (Fig. 4.1), because the spins for the two branches are opposite and therefore the states are orthogonal to each other. Therefore the gapless nature remains, and the Dirac point moves away from $\mathbf{k} = 0$. In the Dirac model, the surface states have the mirror symmetry for any plane normal to the surface. Hence the gapless dispersion appears for any direction of the in-plane Zeeman field. Nevertheless generally in lattice systems, the mirror symmetry are restricted to several special crystallographic planes. In other words, if the Zeeman

R. Takahashi, *Topological States on Interfaces Protected by Symmetry*,
Springer Theses, DOI 10.1007/978-4-431-55534-6_4

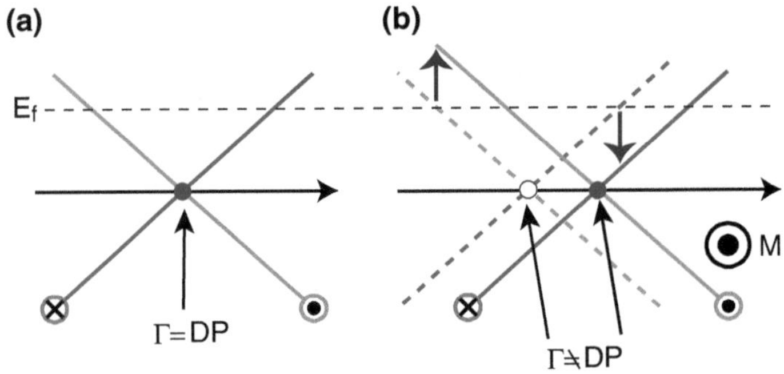

Fig. 4.1 Schematics of the surface dispersion in the mirror plane. In the figure, DP denotes Dirac point, and Γ is at $\mathbf{k} = 0$. **a** The surface dispersion without the Zeeman field. ⊙ and ⊗ attached to the dispersion denote the spin directions normal to the mirror plane, which point to opposite directions. **b** The *solid line* is the surface dispersion with a Zeeman field normal to the mirror plane. The *dotted line* shows the surface dispersion without the Zeeman field. By the in-plane Zeeman field, each state shifts in energy, and gapless states remain

field is not normal to the crystallographic mirror plane, the surface states will become gapped in general.

As an application of the above theory of the gapless states protected by the mirror symmetry, we discuss a Weyl semimetal [1–5]. In Weyl semimetals, there are an even number of bulk Dirac cones without degeneracy. In this cone, the Dirac point is called Weyl node. Around the Weyl node, the effective model is typically expressed as $H_W \sim \pm\mathbf{k}\cdot\boldsymbol{\sigma}$ in the Brillouin zone (BZ), and the dispersion is linear in $\mathbf{k}$. In the Weyl semimetal there are surface states whose Fermi surface forms an arc, connecting the Weyl nodes in the surface-projected BZ. Several works propose how to realize such states [3–5] by breaking the time-reversal symmetry or parity.

A Weyl semimetal phase in a film of TIs is originally proposed by Cho [2]. In this work, the Weyl semimetal phase is realized, by using the effective surface model in the surface of 3DTI in the in-plane Zeeman field. In this dissertation, we use almost the same model for the low-energy model, but we discuss the gapless nature in terms of the mirror symmetry, and we propose a new lattice model to describe the Weyl semimetal phase. We then show that the velocity of the edge states of the Weyl semimetal is modulated when an electric voltage normal to the surface is applied. The change of velocity is caused by broken inversion symmetry.

The system which we propose is a thin slab of TI in a magnetic field (Fig. 4.2). The surface states in TIs are gapped when the thickness of the slab is finite, because the top and bottom surface states are hybridized (Fig. 4.3a). However, as we show later the surface gap closes by magnetization, and the system regains the gapless nature. The purpose of this chapter is to investigate the gapless states and edge states of a thin film of a TI in a Zeeman field. The low-energy Hamiltonian in the two-dimensional (2D) momentum space ($\mathbf{k}$) is written as

$$H(\mathbf{k}) = \left(\frac{v_1+v_2}{2}\tau_z + \frac{v_1-v_2}{2}\right)(\sigma_x k_y - \sigma_y k_x) + \Delta\tau_x - \boldsymbol{\sigma}\cdot\mathbf{m} + g\tau_z. \quad (4.4)$$

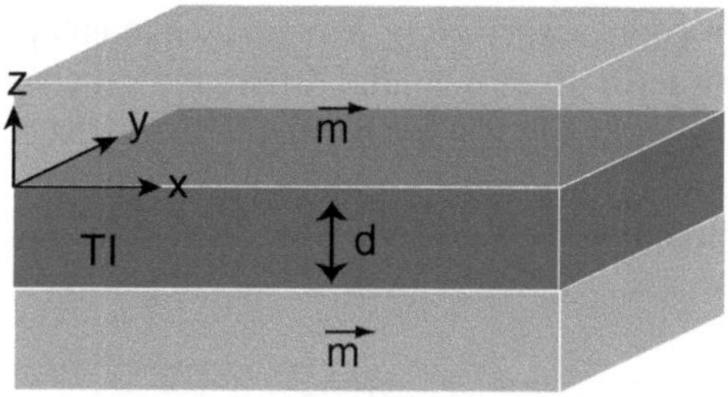

Fig. 4.2 Schematic of the thin slab of a TI sandwiched between magnetic films. **m** denotes the magnetization of the magnet. When the thickness d is small compared with the penetration depth of the surfaces states λ, the *top* and *bottom* surface are hybridized with each other, and the gap due to the finite-size effect cannot be ignored

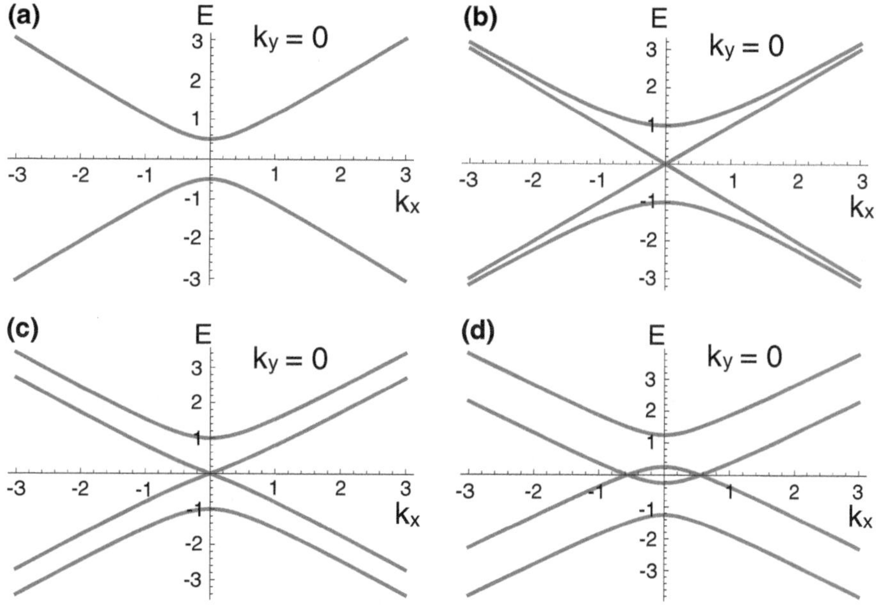

Fig. 4.3 Dispersions for Eq. (4.5). In the results, we employ $\Delta = 0.5$, and $m_x = 0.0$. In **a** the system retains time-reversal symmetry, $m = 0.0$. For (**b, c, d**), time-reversal symmetry is broken by the magnetization, and we show dispersions when they become gapless. **b** $\mathbf{m} = (0, 0, -0.5)$. **c** $\mathbf{m} = (0, -\sqrt{0.125}, -\sqrt{0.125})$. By increasing m, the bands in (**b**) and (**c**) at $E = 0$ split. **d** $\mathbf{m} = (0, -0.75, 0)$. In this case the gapless states remain even when $m > \Delta$, because the magnetization is parallel to the surface

In the right hand side of the above equation, the first term represents the top and bottom surface states, and $v_{1(2)}$ is the Fermi velocity of the top (bottom) surface. We assume that each surface state has a single Dirac cone. The second term describes the finite-size effect, which causes the surface gap $\Delta(>0)$. The third term describes the Zeeman effect. $\boldsymbol{\sigma}$, $\boldsymbol{\tau}$ are Pauli matrices acting on the spin, and the surface (top and bottom) degrees of freedom respectively. We treat the magnetization $\mathbf{m}$ as a tunable parameter. The fourth term gives potential difference g between the two surfaces, which corresponds to an electric voltage normal to the surface, and this term breaks the inversion symmetry.

To simplify the model we first consider the case with $v_1 = v_2$ and $g = 0$. The eigenvalue $E(\mathbf{k})$ satisfies

$$E^2(\mathbf{k}) = \Delta^2 + m^2 + k^2 \pm 2\sqrt{\Delta^2 m^2 + (m_x k_y - m_y k_x)^2}, \tag{4.5}$$

where m, k are absolute values of the magnetization and wavevector, and for simplicity we put $v_{1,2} = 1$. We call the four energies given by Eq. (4.5) in descending order as $E_{i=1,2,3,4}$. The second and third bands E_2, E_3 can be degenerate by tuning parameters. The condition for the two bands to touch is given as

$$(k^2 + \Delta^2 + m^2 - 2m_y'^2)^2 + 4(m_y'^2 - m^2)(\Delta^2 - m_y'^2) = 0, \tag{4.6}$$

where $m_y' = (-m_x k_y + m_y k_x)/k = m_\parallel \sin(\phi - \alpha)$. Here the momentum and magnetization are represented by cylindrical coordinates: $\mathbf{k} = (k\cos\alpha, k\sin\alpha)$ and $\mathbf{m} = (m_\parallel \cos\phi, m_\parallel \sin\phi, m_z)$. We consider two cases for the in-plane and out-of-plane magnetization separately, to search for wavenumbers which satisfy Eq. (4.6).

Out-of-plane magnetization: For $m_z \neq 0$, the solution is $k = 0$ and $m = \Delta$.[1] For $k \neq 0$, to satisfy Eq. (4.6), $\Delta^2 - m_y'^2$ and $\Delta^2 + m^2 - 2m_y'^2$ must be negative, which leads to $\frac{m^2+\Delta^2}{2} < m_y'^2 < \Delta^2$. However there is no solution for the condition. We only have to consider the second and third eigenvalues at $k = 0$ to discuss the gap-closing phenomenon; the energies are $E(0) = \pm(\Delta - m)$. The surface gap becomes smaller with an increase of the magnetization from $m = 0$, and it closes at $m = \Delta$ (Fig. 4.3b, c), and then the Zeeman splitting makes the gap. At $m = \Delta$ the gap closes and the topological phase changes with the change of the Chern number. For $m > \Delta$, the system shows the QH effect, and edge states appear in the system.

In-plane magnetization: For $m \geq \Delta$ and $m_y' = \pm m$, the solution for Eq. (4.6) can be obtained as $k = \pm\sqrt{m^2 - \Delta^2}$. The condition, $m_y' = \pm m$, means that the wavevector for gap closing is perpendicular to the magnetization, $\alpha - \phi = \pm\frac{\pi}{2}$. Namely, for $\mathbf{m} = (0, m, 0)$, for example, the gap-closing points are $(\pm\sqrt{m^2 - \Delta^2}, 0)$ in the momentum space. For the previous case, $m_z \neq 0$, the surface states obtain the gap with the magnetization due to the Zeeman splitting. When the magnetization is in-plane, however, the gapless states remain gapless when the magnetization is sufficiently large ($|m| > \Delta$), and the surface gap due to the hybridization vanishes. There are then two Weyl nodes in the dispersion (Fig. 4.3d). This system is a 2D Weyl semimetal similar to graphene, and the edge states also appear because the mass of the states changes at $k = 0$, $m = 0$.

---

[1] Here we consider the case where $\mathbf{m}$ is not necessarily parallel to z axis.

4.1.1 Weyl Semimetal Under Broken Inversion Symmetry

We have obtained the Weyl semimetal phase in the thin TI by the in-plane Zeeman field. Here by breaking the inversion symmetry, we show that the gapless nature is robust and protected by the mirror symmetry. To break the inversion symmetry, we assume $v_1 \neq v_2$ and $g \neq 0$ in the following discussion. At $k = 0$, the four eigenvalues are $E = \pm\tilde{\Delta} \pm m$, where $\tilde{\Delta} = \sqrt{\Delta^2 + g^2}$. Hence at $\tilde{\Delta} = m$, with changes in the magnetization from zero, the gap closes, and the topological phase changes (Fig. 4.4a, b). Without loss of generality, here we assume the magnetization as $\mathbf{m} = (0, -m, 0)$; namely, the magnetization is along $-y$ direction. At $k = 0$, spins of the four eigenstates are along $\pm y$ directions. For $m < \tilde{\Delta}$ (Fig. 4.4a), the two states in the upper (E_1 and E_2) or the lower (E_3 and E_4) pairs of the bands have opposite direction of spins. At $m = \tilde{\Delta}$, the gap closes, and then for $m > \tilde{\Delta}$ (Fig. 4.4b) the second and third of the four bands are inverted at $k = 0$. When k is far away from $k = 0$ ($v_{1,2}k \gg \Delta, m, g$), the energy spectrum is treated as two single Dirac cones with opposite chiralities; the upper (E_1 and E_2) or the lower (E_3 and E_4) pairs of the bands have opposite direction of spins. For $m > \tilde{\Delta}$, each pair has the same direction of the spins at $k = 0$, but when k is far away from zero the pair must have opposite directions of spins. Thus the second and third of the four energy bands inevitably cross in $k_y = 0$ plane (Fig. 4.4c), and there are two Dirac cones in the BZ. These Weyl nodes are protected by the mirror symmetry, because when the magnetization

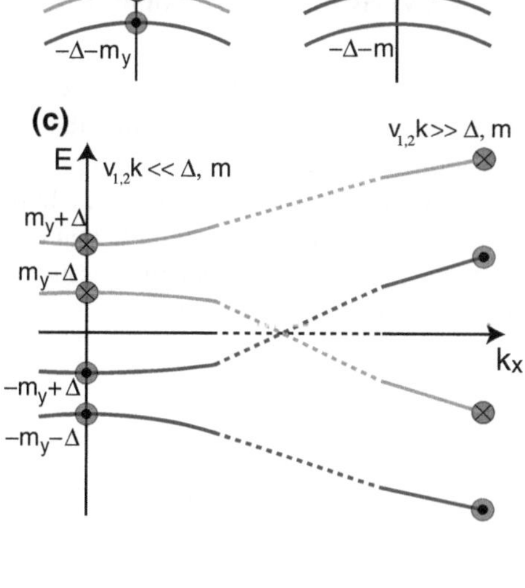

Fig. 4.4 Schematics of the dispersions in $k_y = 0$. When $g \neq 0$, Δ is replaced with $\tilde{\Delta}$. ⊗ (⊙) means that s_y points to + (-) y axis. **a** For $m_y < \Delta$. **b** For $m > \Delta$ and $m_z \neq 0$. **c** For $m_y < \Delta$. Because each state is an exact eigenstate of σ_y and it retains its spins, the two states in the middle cross each other. When the inversion symmetry is broken, the crossing point of the energy shifts

is in-plane, the system has the mirror symmetry with respect to a plane normal to the magnetization, and spins are fixed perpendicular to the $k_y = 0$ plane. Therefore the system becomes a 2D Weyl semimetal, which has two single Dirac cones.

From the above discussion, we have found that to obtain the gapless states, mirror symmetry with respect to the plane normal to the surface is necessary in this mechanism. The difference between the two surface velocities ($v_1 \neq v_2$) or the gate voltage ($g \neq 0$) breaks the inversion symmetry, but they do not violate the mirror symmetry. Therefore the gapless nature is topologically preserved, and the energy where the second and third bands touch shifts due to them. This gives a finite velocity to the edge states because edge states in a Weyl semimetal connect two Weyl nodes. This finite velocity of the edge states is quite different from graphene or any other Weyl semimetals where the edge states have zero velocity. Our result is generic as long as the mirror symmetry exists. The energy shift $\delta E_{0\pm}$ and momenta $\mathbf{k}_{0\pm}$ where the two band touch are derived by Eq. (4.4) as

$$\mathbf{k}_{0\pm} = \pm m\sqrt{\frac{m^2 - \tilde{\Delta}}{m^2 - g^2}}, \quad \delta E_{0\pm} = \pm g\sqrt{\frac{m^2 - \tilde{\Delta}}{m^2 - g^2}}, \tag{4.7}$$

where we assume $v_1 = v_2$ for simplicity. By the above result, the velocity of the edge states is estimated as g/m, proportional to the field g, breaking the inversion symmetry.

In general, it is difficult to observe 2D Weyl semimetals experimentally. However the finite velocity at the edge states can be used for observation of the edge states. By fixing the in-plane magnetization perpendicular to the mirror plane, the conductivity varies as the changes of the gate voltage because the edge current flows in the single direction with the velocity ($\sim g/m$). By measuring the difference of the conductivities between gate voltages g and $-g$, the existence of the Weyl semimetal can be observed.

4.2 Weyl Semimetal Phase in a Lattice Model

The gapless states and the edge states of the 2D Weyl semimetal are the main results of this chapter. We have shown that the magnetization causes the topological phase transition, and their gapless nature is protected by the mirror symmetry. To show how edge states evolve in the slab, we use a lattice model which describes the low-energy Hamiltonian Eq. (4.4). We assume that the system is in a topologically trivial phase due to the hybridization when there is no magnetization. We construct a tight-binding model on the square lattice as:

$$\begin{aligned} H_L = &\frac{i}{4}\sum_{\langle i,j\rangle} c_i^\dagger((v_1+v_2)\tau_z + (v_1 - v_2))[\boldsymbol{\sigma} \times \mathbf{d}_{ij}]_z c_j \\ &+ \sum_{\langle i,j\rangle} c_i^\dagger(2\delta_{ij} - \frac{1}{2}\tau_x)c_j + \sum_i c_i^\dagger(\Delta\tau_x + g\tau_z)c_i - \sum_i c_i^\dagger \tau_0 \boldsymbol{\sigma}\cdot\mathbf{m}c_i, \end{aligned} \tag{4.8}$$

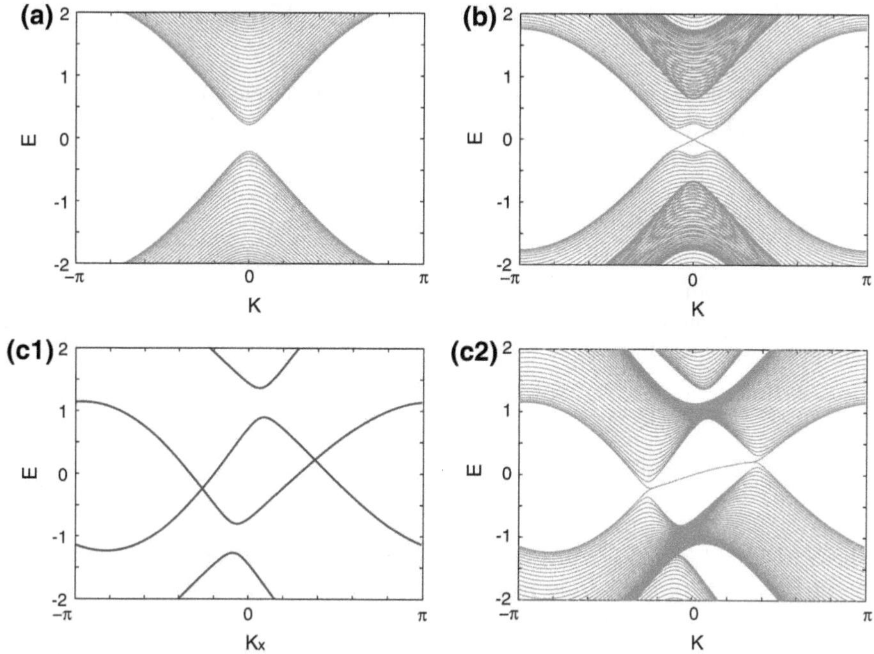

Fig. 4.5 Dispersion of the lattice system Eq. (4.8). The hybridization gap $\Delta = 0.2$, and $m_x = 0$. **a** and **b** show the dispersions in the ribbon geometry with inversion symmetry ($v_1 = v_2 = 1.0$, and $g = 0$). **a** $m_y = 0.1$ and $m_z = 0$. The system is in topologically trivial phase. **b** $m_y = 0.4$ and $m_z = 0.2$. The system shows the QH phase. In (**c1**) and (**c2**) we employ the parameters as $g = 0.3$, $m_y = 1.08$, $m_z = 0$, $v_1 = 0.8$, $v_2 = 1.2$ where the inversion symmetry is broken. **c1** Shows the bulk bands at $k_y = 0$, and **c2** shows energy bands in the ribbon geometry

where c_i ($c_i^\dagger$) is the annihilation (creation) operator of the electron at ith site. $\sum_{\langle i,j\rangle}$ denotes the sum over the nearest-neighbor pairs. $\mathbf{d}_{ij}$ is a vector from jth site to ith site. There are four components due to the surface and spin degree of freedom at each site in the 2D lattice system, and Δ represents a mass due to the finite thickness of the slabs. We show dispersions for edge states by imposing the open boundary condition. Dispersions of the lattice model are shown in Fig. 4.5. The gap is fixed as $\Delta = 0.2$ for the results **a**,**b**,**c**. For **a** and **b** the system has the inversion symmetry ($g = 0$, $v_1 = v_2$); the magnetizations are $\mathbf{m} = (-0.1, 0, 0)$ for **a**, and $\mathbf{m} = (-0.4, 0, -0.2)$ for **b**. In Fig. 4.5**a**, there is no edge state because $m < \Delta$. On the other hand there appear edge states due to the QH effect for $m > \Delta$ and $m_z \neq 0$ (Fig. 4.5b). In **c1** and **c2**, the inversion symmetry is broken ($g = 0.3$, $v_1 = 0.8$, $v_2 = 1.2$), and the magnetization is employed as $\mathbf{m} = (-1.08, 0, 0)$ which satisfies $m > \Delta$ and $m_z = 0$. Figure 4.5**c1**, **c2** shows energy spectra of the bulk at $k_y = 0$, in the ribbon geometry, respectively. In the results we set the parameters as $m > \Delta$ and $m_z = 0$. The system is a 2D Weyl semimetal, and the edge states have a finite velocity originating from the broken inversion symmetry, in agreement with the previous section.

4.3 Estimation of the Magnetization at the Phase Transition

Here we estimate the magnetization when the topological transition occurs, in Bi_2Se_3 known as a 3DTI. To simplify the model, we assume that the system has the inversion symmetry ($g = 0$ and $v_1 = v_2$), and the surface gap Δ is described approximately as:

$$\Delta(d) = \Delta_0 e^{-\frac{d}{\lambda}}, \tag{4.9}$$

where Δ_0 is a half of the bulk gap.

To introduce the Zeeman field, we follow Ref. [6] for the calculation. The Zeeman field **m** is represented as

$$\begin{aligned} m_z &= \frac{\mu_B}{2} g_{xz} B_z, \\ m_p &= \frac{\mu_B}{2} g_{sp} B_p, \end{aligned} \tag{4.10}$$

where $B_{z(p)}$ and $g_{xz(sp)}$ is a magnetic fields and g-factor, perpendicular (parallel) to the surface, and μ_B is the Bohr magneton. The values of g-factors are theoretically proposed as shown in Table 4.1 [6].

The magnetic field B_T when the transition occur is expressed as $m = \Delta$, i.e. $B_T = \frac{2\Delta(d)}{\mu_B g}$.[2] We plot B_T as a function of the number of quintuple layers (QL$\sim$ 1 nm) for in Bi_2Se_3 in Fig. 4.6. B_{Tp} and B_{Tz} are in-plane and out-of-plane magnetic fields at which the phase transition occurs. We employ the parameters for Bi_2Se_3:

Table 4.1 g-factors of the surface states in Bi_2Se_3, Bi_2Te_3, and Sb_2Te_3 from Ref. [6]

	Bi_2Se_3	Bi_2Te_3	Sb_2Te_3
g_{sp}	1.01	0.838	8.4835
g_{sz}	−8.43	−17.4	−2.09

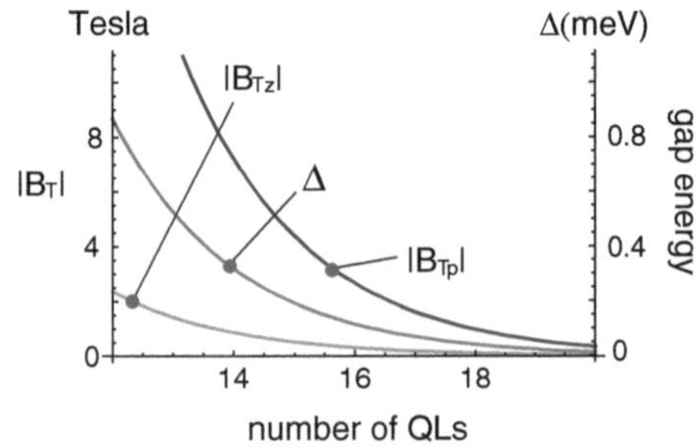

Fig. 4.6 Plots of B_{Tp}, B_{Tz} and the gap due to the finite size effect Δ as functions of the number of QLs in Bi_2Se_3

[2] When the gate voltage V normal to the film is applied, B_T has the form $B_T = \frac{2}{\mu_B g}\sqrt{\Delta^2(d) + V^2}$.

$\lambda = 2$ QL [7] and $\Delta_0 = 0.35$ eV. When $d = 15$ QL, the those transitional magnetic fields at the phase transition are $|B_{Tp}| = 4.38$T, $|B_{Tz}| = 0.52$T and the gap is about 0.19 meV.

References

1. M.Z. Hasan, C.L. Kane, Rev. Mod. Phys. **82**, 3045 (2010)
2. G.Y. Cho, arXiv:1110.1939v2
3. X. Wan et al., Phys. Rev. B **83**, 205101 (2011)
4. A.A. Burkov, L. Balents, Phys. Rev. Lett. **107**, 127205 (2011)
5. A.A. Burkov et al., Phys. Rev. B **84**, 235126 (2011)
6. C.-X. Liu et al., Phys. Rev. B **82**, 045122 (2010)
7. W. Zhang et al., New J. Phys. **12**, 065013 (2010)

Chapter 5
Summary and Outlook

In the present dissertation, first we studied refraction phenomena at a junction between two topological insulator surfaces with different velocities in Chap. 3. A Snell's law for the surface refractive phenomena is obtained by the energy and momentum conservation. We calculated refraction coefficients where the velocity depends on the position, and derived the transmittance and reflectance. We found that the wavefunction is not continuous by the current conservation at the interface. Unlike optics, the surface electronic wave is perfectly transmitted when the incident wavevector is perpendicular to the boundary. This is due to the prohibited backscattering in the topological insulator surface; in the surface states, spins for the wavenumbers $\mathbf{k}$ and $-\mathbf{k}$ are opposite. When the velocities of the topological insulator surface states have different signs, we showed that the gapless states appear on the interface. In this situation, both the transmission and reflection for the normal incidence is prohibited due to the spin conservation, when we consider the refraction problem as in the previous case. The dispersions of interface states are obtained by using the low-energy Hamiltonian. There appear two Dirac points on the mirror plane, which reflect the mirror symmetry. In addition, when the system has the C_3 rotational symmetry there are six Dirac cones which reflect the rotational symmetry. These six Dirac cones are interpreted as follows; there are three mirror planes due to the C_3 symmetry, and on each mirror plane there are two Dirac points. The existence of the gapless states is generally shown by using the mirror Chern number and is topologically protected by the mirror symmetry. The mirror Chern number is well defined when the system is mirror symmetric, and it corresponds to the chirality of the spins on the Fermi surface. Furthermore, the chirality coincides with the sign of the Dirac velocity. Therefore the sign of the velocity is topologically classified by the mirror Chern number, and it is natural that their difference gives the boundary states between two topological insulators.

Next we studied the mirror Chern number in the Fu-Kane-Mele tight-binding model on the diamond lattice. The calculations of the mirror Chern number are performed by two methods, which generate identical results; one is a method to calculate the change of the Chern number in the parametric space at a phase transition, and the other is a numerical integration of the Berry curvature on the mirror plane.

R. Takahashi, *Topological States on Interfaces Protected by Symmetry*,
Springer Theses, DOI 10.1007/978-4-431-55534-6_5

In the Fu-Kane-Mele model, we found that the sign of the spin-orbit interaction, $\text{sign}(\lambda)$, corresponds to the mirror Chern number; when $\text{sign}(\lambda) = +(-)$, the mirror Chern number becomes $-1(+1)$. Furthermore we showed the gapless interface states between the two regions in the Fu-Kane-Mele tight-binding model, whose spin-orbit interaction is opposite in sign. We showed that two Dirac points appear on the mirror plane. When the hopping integral preserves the C_3 symmetry, there appear six Dirac cones, which well agrees with the low-energy model.

From the result of the mirror Chern number in the Fu-Kane-Mele model we found that the sign of the spin-orbit interaction corresponds to the chirality. Thus, we can say that the chirality of the surface states will be determined by the sign of the spin-orbit interaction in general systems, because the spin-filtered phenomena in topological insulators originate from the spin-orbit interaction. If one would like to further investigate the relation between the chirality and sign of the spin-orbit interaction, an ab-initio calculation for the surface dispersion in a topological insulator will be necessary. Through the calculation, one will need to separate the spin-orbit interaction from the Hamiltonian, and to analyze the relation between its sign and the mirror Chern number.

To realize our gapless interface states with an even number of Dirac cones experimentally, two topological insulators with opposite chiralities are necessary and mirror symmetry normal to the interface is required in common. All typical topological insulators have the same chirality. However a recent experiment has reported $Bi_4Se_{2.6}S_{0.4}$ as a topological insulator with the opposite chirality from the typical one, which is based on natural superlattice of alternating Bi_2 layers and Bi_2Se_3 layers, and sulfur (S) is added in Bi_2Se_3 layers. If $Bi_4Se_{2.6}S_{0.4}$ has good interfaces with other topological insulators, especially Bi_2Se_3, there is a possibility to realize the gapless interface states, although the conditions seem to be restricted. These interface states are similar to topological crystalline insulators, which also have gapless states with an even number of Dirac cones on the surface, protected by the mirror symmetry. The topological crystalline insulators can be weak topological insulators, and they have been observed experimentally.

In addition, we gave another kind of interface metallic states originating from new symmetries called interfacial symmetries in junction systems: interfacial particle-hole symmetry (IPHS) and interfacial time-reversal symmetry (ITHS). In these junction systems with interfacial symmetries, the whole system is invariant under the particle-hole or time-reversal transformation, while the respective regions on either side of the junction are not necessarily invariant by the transformation. As an example, we showed the Fermi loop in the interface between two regions of the FKM models with opposite signs of the spin-orbit interaction parameters, having the IPHS. The interface Fermi loop is guaranteed by the reality of the Pfaffian of a skew-symmetric matrix: $\text{pf}(\tilde{U}H)$ where $\tilde{U}$ consists of the space inversion and the particle-hole transformation. We found that the Fermi loop is a boundary between the positive and negative regions of $\text{pf}(\tilde{U}H)$. This presence of the Fermi loop is shown for general systems with the IPHS. On the other hand, the ITRS leads to double degeneracy for every state in junction systems. We showed the Fermi loops in the π-junction interface between the two Rashba systems as another example of the IPHS. This implies our theory on the interfacial symmetry can explain emergence of boundary states between trivial superconducting phases.

In Chap. 4 we discussed the robustness of a surface Dirac cone against in-plane magnetic fields, caused by mirror symmetry. When the system has the mirror symmetry, the surface Dirac cone remains gapless by the Zeeman field perpendicular to the mirror plane. Because the mirror symmetry fixes the spin on the mirror plane, the spin on the mirror plane does not change if the magnetic field is applied perpendicular to the mirror plane. In addition, we studied edge states in a thin topological insulator subjected to a Zeeman field. When the magnitude of the magnetization is larger than the energy scale of the finite-size effect, the topological phase changes; when the magnetization is out-of-plane the system becomes the quantum Hall phase, and when the magnetization is in-plane it becomes the Weyl semimetal phase. We found that in the Weyl semimetal phase the edge states have a finite velocity when the inversion symmetry is broken. This theory is also verified in a lattice model. In the lattice model, we showed dispersions in a ribbon geometry in one dimension, and dispersions of the edge states in the Weyl semimetal phase. Moreover we estimated the magnetization for the phase transition in Bi_2Se_3; when the thickness of the system is about $d = 15$ quintuple-layers ($\sim$15 nm), the gap by the finite-size effect is about 1.9 meV, and the transition magnetic fields are about 0.52 T for the quantum Hall phase, and about 4.38 T for the Weyl semimetal phase.

In general, it is difficult to observe two-dimensional Weyl semimetals by the angle-resolved photoemission spectroscopy because the edge is confined in one dimension. The finite velocity of the edge states can be useul for the observation. With the in-plane magnetization fixed perpendicular to the mirror plane, the conductivity varies as the gate voltage changes because the edge state is one-way. By measuring the difference of the conductivities between gate voltages g and $-g$ and extracting the edge contribution, the existence of the Weyl semimetal can be observed. Furthermore our estimation of the magnetization shows realistic values for experiments to realize the Weyl semimetal. We note that one needs to have good resolution in the measurement for the observation under the small magnetization. The transition magnetization becomes smaller with the increase in the thickness and, the thinner the slab gets, the larger the gap becomes. Therefore, it is easy to realize the Weyl semimetal phase in a thick slab which has a small gap. However, when the gap is small there are only a few edge states. In addition, unlike topological insulators, there are bulk states at the Fermi energy in the Weyl semimetal. Therefore, it will be difficult to observe the edge conductance in the weak magnetization.

Researches on topological insulators have been intensively done recently. In particular, recently the effect of electron-electron interaction on topological insulators has been actively studied. This tendency is historically similar to researches on the quantum Hall effect; researchers have been interested in fractional topological insulators. Additionally, it has been of particular interest to find new topological orders induced by interaction. In the dissertation we proposed several scenarios for new types of surface/interface states coming from topology based on single-particle physics. Particularly, we showed that mirror symmetry plays a crucial role in this gapless nature. Researches on topological insulators in terms of such a symmetry are only a few, and will be likely to increase. Moreover there are a few researches on topological phenomena in terms of interfaces originating from combined symmetry like that in Sect. 3.4. Therefore the results presented in this dissertation will stimulate future research if they are realized. Realizations of our proposals remain as future works. In the two subjects in the dissertation, we mentioned how to realize them experimentally: the gapless interface states protected by mirror symmetry and the Weyl semimetal in thin topological insulators. At the same time, some of the theoretical results in this dissertation would lead us to new classes of topological order and emergent topological phenomena.

Chapter 6
Properties of the Chern Numbers

6.1 Chern Number

In this section we explain the basic properties of the Chern number (the TKNN number). We focus on one band which is completely filled, and show that the value ν in Eq. (2.6) is an integer. By Eq. (2.6), we have

$$\begin{aligned}\nu &= \frac{i}{2\pi}\int_{\mathrm{BZ}} d\mathbf{k}(\langle\partial_x u(\mathbf{k})|\partial_y u(\mathbf{k})\rangle - \langle\partial_y u(\mathbf{k})|\partial_x u(\mathbf{k})\rangle) \\ &= \frac{1}{2\pi}\int_{\mathrm{BZ}} d\mathbf{k}\nabla\times\mathbf{a}(\mathbf{k})\cdot\tilde{z} \\ &= \frac{1}{2\pi}\oint_{\mathrm{BZ}} d\mathbf{k}\cdot\mathbf{a}(\mathbf{k}),\end{aligned} \tag{6.1}$$

where $\mathbf{a}(\mathbf{k}) = i\langle u(\mathbf{k})|\nabla u(\mathbf{k})\rangle$ is the Berry connection, Stokes's theorem is used, $\tilde{z}$ is (0, 0, 1), and BZ denotes the Brillouin zone. For simplicity, we consider the case where the two primitive reciprocal vectors which are orthogonal to each other, $\mathbf{G}_x = (G_{0x}, 0)$ and $\mathbf{G}_y = (0, G_{0y})$. Extension to generic cases is straightforward. Then, we have

$$\begin{aligned}\nu = &\frac{i}{2\pi}\int_0^{G_{0x}} dk_x\langle u(k_x,0)|\partial_x u(k_x,0)\rangle + \frac{i}{2\pi}\int_0^{G_{0y}} dk_y\langle u(G_{0x},k_y)|\partial_y u(G_{0x},k_y)\rangle \\ &+\frac{i}{2\pi}\int_{G_{0x}}^0 dk_x\langle u(k_x,G_{0y})|\partial_x u(k_x,G_{0y})\rangle + \frac{i}{2\pi}\int_{G_{0y}}^0 dk_y\langle u(0,k_y)|\partial_y u(0,k_y)\rangle,\end{aligned} \tag{6.2}$$

and the BZ of the system is

$$0 \le k_x \le G_{0x},\ 0 \le k_y \le G_{0y}. \tag{6.3}$$

R. Takahashi, *Topological States on Interfaces Protected by Symmetry*,
Springer Theses, DOI 10.1007/978-4-431-55534-6_6

Because states at $\mathbf{k}$ and $\mathbf{k}+\mathbf{G}_{x,y}$ are the same, we can write

$$\begin{cases} |u(G_{0x}, k_y)\rangle = \mathbf{e}^{i\chi_1(k_y)}|u(0, k_y)\rangle, \\ |u(k_x, G_{0y})\rangle = \mathbf{e}^{i\chi_2(k_x)}|u(k_x, 0)\rangle, \end{cases} \tag{6.4}$$

where $\chi_{1,2}$ are real functions. Therefore ν has the form

$$\begin{aligned} \nu = & \frac{i}{2\pi}\int_0^{G_{0x}} \mathrm{d}k_x \langle u(k_x, 0)|\partial_x u(k_x, 0)\rangle \\ & + \frac{i}{2\pi}\int_0^{G_{0y}} \mathrm{d}k_y (\langle u(0, k_y)|\partial_y u(0, k_y)\rangle + i\partial_y \chi_1(k_y)) \\ & + \frac{i}{2\pi}\int_{G_{0x}}^{0} \mathrm{d}k_x (\langle u(k_x, G_{0y})|\partial_x u(k_x, G_{0y})\rangle + i\partial_x \chi_2(k_x)) \\ & + \frac{i}{2\pi}\int_{G_{0y}}^{0} \mathrm{d}k_y \langle u(0, k_y)|\partial_y u(0, k_y)\rangle \\ = & \frac{1}{2\pi}[\chi_1(G_{0y}) - \chi_1(0) - \chi_2(G_{0x}) + \chi_2(0)]. \end{aligned} \tag{6.5}$$

Here by the relation in Eq. (6.4), the phases $\chi_{1,2}$ satisfies the equations

$$|u(G_{0x}, G_{0y})\rangle = \mathbf{e}^{i\chi_1(G_{0y})+i\chi_2(0)}|u(0,0)\rangle = \mathbf{e}^{i\chi_1(0)+i\chi_2(G_{0x})}|u(0,0)\rangle, \tag{6.6}$$

namely,

$$\chi_1(G_{0y}) + \chi_2(0) - \chi_1(0) - \chi_2(G_{0_x}) = 2\pi n, \tag{6.7}$$

where n is integer. Therefore ν is integer, and this is the Chern number also called TKNN integer.

6.1.1 Berry Curvature with Time-Reversal Symmetry

Here, we derive the relation given in Eq. (2.47). We denote the wavefunctions as $|u^{\alpha,\beta}(\mathbf{k})\rangle$ where α and β are band indices transformed to each other through the time-reversal operator Θ. The Berry connections, $\mathbf{a}^{\alpha,\beta}(\mathbf{k}) = i\langle u^{\alpha,\beta}(\mathbf{k})|\nabla u^{\alpha,\beta}(\mathbf{k})\rangle$, satisfy

$$\begin{aligned} \mathbf{a}^{\alpha}(-\mathbf{k}) &= -i\langle u^{\alpha}(-\mathbf{k})|\nabla u^{\alpha}(-\mathbf{k})\rangle \\ &= -i\langle \nabla\Theta u^{\alpha}(-\mathbf{k})|\Theta u^{\alpha}(-\mathbf{k})\rangle \\ &= i\langle \Theta u^{\alpha}(-\mathbf{k})|\nabla\Theta u^{\alpha}(-\mathbf{k})\rangle \\ &= \mathbf{a}^{\beta}(\mathbf{k}) + i\nabla\chi(\mathbf{k}), \end{aligned} \tag{6.8}$$

where we put the wavefunctions as

$$|u^{\beta}(\mathbf{k})\rangle = e^{i\chi(\mathbf{k})}|\Theta u^{\alpha}(-\mathbf{k})\rangle, \tag{6.9}$$

and $\chi(\mathbf{k})$ is real. Thus, the Berry curvatures $\mathbf{b}^{\alpha,\beta}$ satisfy the relation

$$\begin{aligned}\mathbf{b}^{\beta}(\mathbf{k}) &= \nabla \times \mathbf{a}^{\beta}(\mathbf{k}) \\ &= -\mathbf{b}^{\alpha}(-\mathbf{k}).\end{aligned} \tag{6.10}$$

Therefore, in total, the Berry curvature $\mathbf{b}(\mathbf{k}) = \mathbf{b}^{\alpha}(\mathbf{k}) + \mathbf{b}^{\beta}(\mathbf{k})$ satisfies

$$\mathbf{b}(\mathbf{k}) = -\mathbf{b}(-\mathbf{k}), \tag{6.11}$$

because of the time-reversal symmetry.

6.2 Kramers Theorem

The time-reversal symmetry plays a crucial role in topological insulators as explained in Sect. 2.2.1. Here we briefly review a consequence of the time-reversal symmetry in quantum mechanics. Let us write an eigenvalue and its eigenvector as $E_{\mathbf{k}}$ and $|u(\mathbf{k})\rangle$ of the time-invariant Hamiltonian H. Then the Schrödinger equation is given as

$$H(\mathbf{k})|u(\mathbf{k})\rangle = E_{\mathbf{k}}|u(\mathbf{k})\rangle, \tag{6.12}$$

and because the system has time-reversal symmetry, we have

$$H(-\mathbf{k})\Theta|u(\mathbf{k})\rangle = E_{\mathbf{k}}\Theta|u(\mathbf{k})\rangle. \tag{6.13}$$

Therefore, the states $|u(\mathbf{k})\rangle$ and $\Theta|u(\mathbf{k})\rangle$ are degenerate. In addition, they are orthogonal to each other, because

$$\langle u(\mathbf{k})|\Theta u(\mathbf{k})\rangle = \langle \Theta^2 u(\mathbf{k})|\Theta u(\mathbf{k})\rangle = -\langle u(\mathbf{k})|\Theta u(\mathbf{k})\rangle. \tag{6.14}$$

Thus we conclude $\langle u(\mathbf{k})|\Theta u(\mathbf{k})\rangle = 0$[1]; it is a well-known result from the Kramers theorem.

[1] For states $|u\rangle$ and $|v\rangle$, their product satisfies $\langle u|v\rangle = \langle \Theta v|\Theta u\rangle$.

6.3 Berry Curvature and Chern Number

We derive some formulas for the Berry phase for a 2×2 Hamiltonian, depending on two parameters, R_1, and R_2 [1]. The following results are used in Sect. 3.3.2. For the Hamiltonian $H(\mathbf{R} = (R_1, R_2))$, the Berry curvature is given as

$$b^n_{R_1,R_2} = i\left[\langle \partial_{R_1} n(\mathbf{R})|\partial_{R_2} n(\mathbf{R})\rangle - \langle \partial_{R_2} n(\mathbf{R})|\partial_{R_1} n(\mathbf{R})\rangle\right] \tag{6.15}$$

$$= i \sum_{n' \neq n} \frac{\langle n(\mathbf{R})|\partial_{R_1} H|n'(\mathbf{R})\rangle\langle n'(\mathbf{R})|\partial_{R_2} H|n(\mathbf{R})\rangle}{(E_n(\mathbf{R}) - E_{n'}(\mathbf{R}))^2} + \text{c.c.}, \tag{6.16}$$

where $|n(\mathbf{R})\rangle$ is the nth eigenfunction, E_n is the eigenvalue, and we used

$$\langle \partial_{R_i} n(\mathbf{R})|n'(\mathbf{R})\rangle = \frac{\langle n(\mathbf{R})|\partial_{R_i} H|n'(\mathbf{R})\rangle}{E_n(\mathbf{R}) - E_{n'}(\mathbf{R})}, \tag{6.17}$$

for $n \neq n'$. Without losing generality, the 2×2 Hamiltonian is expressed as

$$\begin{aligned} H &= \mathbf{d}(\mathbf{R}) \cdot \boldsymbol{\sigma} \\ &= \begin{pmatrix} d_z & d_x - id_y \\ d_x + id_y & -d_z \end{pmatrix}, \end{aligned} \tag{6.18}$$

where the trace part is omitted because it does not affect the Berry curvature. The eigenvalues are $\pm d$, where $d^2 = d_i d_i$ and the eigenstates are given as

$$|+\rangle = \frac{1}{\sqrt{2d(d-d_z)}}\begin{pmatrix} d_x - id_y \\ d - d_z \end{pmatrix}, \quad |-\rangle = \frac{1}{\sqrt{2d(d+d_z)}}\begin{pmatrix} -d_x + id_y \\ d + d_z \end{pmatrix}. \tag{6.19}$$

The Berry curvature has the form

$$\begin{aligned} b^+_{R_1,R_2}(\mathbf{R}) &= i\frac{\langle +(\mathbf{R})|\partial_{R_1} d_i\sigma_i| - (\mathbf{R})\rangle\langle -(\mathbf{R})|\partial_{R_2} d_j\sigma_j| + (\mathbf{R})\rangle - (c.c.)}{(E_+(\mathbf{R}) - E_-(\mathbf{R}))^2} \\ &= i\frac{\partial_{R_1} d_i \partial_{R_2} d_j}{4d^2}\left[\langle +|\sigma_i|-\rangle\langle -|\sigma_j|+\rangle - (c.c.)\right] \\ &= -\frac{\partial_{R_1} d_i \partial_{R_2} d_j}{2d^2}\Im\left[\langle +|\sigma_i|-\rangle\langle -|\sigma_j|+\rangle\right], \end{aligned} \tag{6.20}$$

and we have $b^+_{R_1,R_2}(\mathbf{R}) = -b^-_{R_1,R_2}(\mathbf{R})$. Using

$$\begin{cases} \langle +|\sigma_x|-\rangle = \frac{1}{d\sqrt{(d^2-d_z^2)}}(d_x d_z + id_y d), \\ \langle +|\sigma_y|-\rangle = \frac{1}{d\sqrt{(d^2-d_z^2)}}(d_y d_z - id_x d), \\ \langle +|\sigma_z|-\rangle = \frac{1}{d\sqrt{(d^2-d_z^2)}}(-d^2 + d_z^2), \end{cases} \tag{6.21}$$

we have

$$\begin{cases} \langle +|\sigma_x|-\rangle\langle -|\sigma_y|+\rangle = \frac{1}{d^2(d^2-d_z^2)}(d_x d_y(-d^2+d_z^2)+id_z d(d^2-d_z^2)), \\ \langle +|\sigma_y|-\rangle\langle -|\sigma_z|+\rangle = \frac{1}{d^2(d^2-d_z^2)}(d_y d_z(-d^2+d_z^2)+id_x d(d^2-d_z^2)), \\ \langle +|\sigma_z|-\rangle\langle -|\sigma_x|+\rangle = \frac{1}{d^2(d^2-d_z^2)}(d_z d_x(-d^2+d_z^2)+id_y d(d^2-d_z^2)). \end{cases} \tag{6.22}$$

From the above relations, we obtain the Berry curvature as

$$b^{\pm}_{R_1,R_2}(\mathbf{R}) = \mp\frac{1}{2d^3}\varepsilon_{ijk}\partial_{R_1}d_i\partial_{R_2}d_j d_k. \tag{6.23}$$

When the curvature is integrated over a closed surface $\mathscr{S}$, the Berry phase $\gamma_\pm$ can be expressed as

$$\begin{aligned} \gamma_\pm &= \mp\int_{\mathscr{S}} \mathrm{d}S\frac{1}{2d^3}\mathbf{d}\cdot\frac{\partial\mathbf{d}}{\partial R_1}\times\frac{\partial\mathbf{d}}{\partial R_2} \\ &= \mp\frac{1}{2}\int_{\mathscr{S}} \mathrm{d}S\,\hat{\mathbf{d}}\cdot\frac{\partial\hat{\mathbf{d}}}{\partial R_1}\times\frac{\partial\hat{\mathbf{d}}}{\partial R_2}, \end{aligned} \tag{6.24}$$

where $\hat{\mathbf{d}} \equiv \frac{\mathbf{d}}{|\mathbf{d}|}$.

6.3.1 Change of the Chern Number

When some parameter traverses across a gap-closing point by tuning physical quantities, the Chern number discretely jumps at the gap-closing point. The change of the Chern number corresponds to counting monopoles passed by the parametric space. Here, we show a derivation of the difference of the Chern numbers in a two dimensional space (k_1, k_2) across the gap-closing point. We assume that the Hamiltonian depends on a set of parameters $\mathbf{k} = (k_1, k_2, k_3)$. k_3 is a physical parameter, which can be controlled, and the gap-closing point is $\mathbf{k} = \mathbf{k}^0$. With a positive infinitesimal δ, the difference of the Chern number δn is expressed as

$$\delta n = \frac{1}{2\pi}\int_{BZ}\mathrm{d}k_1\mathrm{d}k_2 b_{k_1k_2}(k_3^0+\delta) - \frac{1}{2\pi}\int_{BZ}\mathrm{d}k_1\mathrm{d}k_2 b_{k_1k_2}(k_3^0-\delta), \tag{6.25}$$

where we define the gauge field $a_i = i\langle n(\mathbf{k})|\partial_i n(\mathbf{k})\rangle$ and its corresponding curvature $\mathbf{b} = \nabla\times\mathbf{a}$. Then, Eq. (6.25) is rewritten as

$$\delta n = \frac{1}{2\pi}\int_{S_{\mathrm{BZ}}}\mathrm{d}\mathbf{S}\cdot\mathbf{b} \tag{6.26}$$

where $S_{\rm BZ}$ is a surface consisting of the BZ at $k_3 = k_3^0 \pm \delta$. by using $\nabla \cdot \mathbf{b} = 0$ except for $\mathbf{k} = \mathbf{k^0}$, $S_{\rm BZ}$ can be reduced to a sphere $\mathscr{S}$ which encircles $\mathbf{k}^0$ and has a small radius in the integral Eq. (6.26). Then the 2×2 Hamiltonian can be expanded in terms of $\mathbf{k}$ as

$$H = \sigma_i \alpha_{ij} (k_j - k_j^0), \tag{6.27}$$

where α_{ij} are constants. From Eq. (6.25), δn has the form

$$\begin{aligned} \delta n &= \frac{1}{2\pi} \int_V \mathrm{d}V \ \nabla \cdot \mathbf{b} \\ &= \frac{1}{2\pi} \int_V \mathrm{d}V \varepsilon_{ijk} \partial_i \partial_j a_k, \end{aligned} \tag{6.28}$$

where V is the region surrounded by S. To simplify the above equation, we transform variables from $\mathbf{k}$ to $\mathbf{d}$ as $d_i = \alpha_{ij}(k_j - k_j^0)$. Then the integrand has the form

$$\varepsilon_{ijk} \partial_i \partial_j a_k = \varepsilon_{ijk} \alpha_{li} \alpha_{hj} \alpha_{mk} \bar{\partial}_l \bar{\partial}_h a_m, \tag{6.29}$$

where $\bar{\partial}_i = \frac{\partial}{\partial d_i}$. Only when $h \neq l \neq m \neq h$, the above r.h.s. is nonzero. Therefore (l, h, m) is a permutation of (i, j, k). In addition, the permutation of (l, h, m) is divided into even or odd permutations from (i, j, k). Then the r.h.s. is expanded as

$$\begin{aligned} &\varepsilon_{ijk} \alpha_{li} \alpha_{hj} \alpha_{mk} \bar{\partial}_l \bar{\partial}_h \bar{a}_m \\ &\quad = \varepsilon_{ijk} (\bar{\partial}_i \bar{\partial}_{i+1} \bar{a}_{i+2} + \alpha_{i+1i} \alpha_{i+2j} \alpha_{ik} \bar{\partial}_{i+1} \bar{\partial}_{i+2} \bar{a}_i + \alpha_{i+2i} \alpha_{ij} \alpha_{i+1k} \bar{\partial}_{i+2} \bar{\partial}_i \bar{a}_{i+1} \\ &\qquad + \alpha_{ii} \alpha_{i+2j} \alpha_{i+1k} \bar{\partial}_i \bar{\partial}_{i+2} \bar{a}_{i+1} + \alpha_{i+2i} \alpha_{i+1j} \alpha_{ik} \bar{\partial}_{i+2} \bar{\partial}_{i+1} \bar{a}_i \\ &\qquad + \alpha_{i+1i} \alpha_{ij} \alpha_{i+2k} \bar{\partial}_{i+1} \bar{\partial}_i \bar{a}_{i+2}) \\ &\quad = \det[\alpha] \varepsilon_{ijk} \bar{\partial}_i \bar{\partial}_j \bar{a}_k. \end{aligned} \tag{6.30}$$

Therefore we can express Eq. (6.25) in terms of $\mathbf{d}$ as

$$\delta n = \mathrm{sgn}[\det(\alpha)] \frac{1}{2\pi} \int_V \mathrm{d}\bar{V} \varepsilon_{ijk} \bar{\partial}_i \bar{\partial}_j \bar{a}_k. \tag{6.31}$$

Expressing $\mathbf{d}$ by its polar angle θ and azimuthal angle ϕ, $\mathbf{d} = d(\sin\theta \cos\phi, \sin\theta \sin\phi, \cos\theta)$, we have the eigenvectors as

$$|+\rangle = \begin{pmatrix} \cos\frac{\theta}{2} \mathrm{e}^{-i\phi} \\ \sin\frac{\theta}{2} \end{pmatrix}, \ |-\rangle = \begin{pmatrix} -\sin\frac{\theta}{2} \mathrm{e}^{-i\phi} \\ \cos\frac{\theta}{2} \end{pmatrix}. \tag{6.32}$$

For the lower energy states, by using Eq. (6.24), we derive δn as

$$\begin{aligned}\delta n &= \mathrm{sign}[\det(\alpha)]\frac{1}{4\pi}\int_0^{\pi}\int_0^{2\pi} \mathrm{d}\theta \mathrm{d}\phi \sin\theta \\ &= \mathrm{sign}[\det(\alpha)].\end{aligned} \tag{6.33}$$

Reference

1. M.V. Berry, Proc. R. Soc. Lond. A **392**, 45 (1984)

Chapter 7
Calculation for the Interface Fermi Loops

7.1 Classification of Interfacial Systems

We consider an interface Hamiltonian given as

$$H_{\mathbf{k}_\parallel} = \begin{pmatrix} H_{\alpha,\mathbf{k}_\parallel} & V_{\mathbf{k}_\parallel} \\ V_{\mathbf{k}_\parallel}^\dagger & H_{\beta,\mathbf{k}_\parallel} \end{pmatrix}, \tag{7.1}$$

where $\mathbf{k}_\parallel$ is the wavevector perpendicular to the interface between the two regions α and β considered in Sect. 3.4. In the respective bulk systems, the pair of Hamiltonians in both regions is connected by a unitary operator U,

$$H_{\alpha(\beta),\mathbf{k}} = \varepsilon_C U H^t_{\beta(\alpha),-\mathbf{k}} U^{-1}, \tag{7.2}$$

with $\varepsilon_C = \pm 1$, where $\mathbf{k}$ is the three-dimensional wavevector. The hybridization V between the two regions is assumed to satisfy the condition,

$$\tilde{P} U V_{\mathbf{k}_\parallel} (\tilde{P} U)^{-1} = \varepsilon_C V^t_{\mathbf{k}_\parallel}, \tag{7.3}$$

where $\tilde{P}$ inverts the order of stacking along the surface normal. The Hamiltonian satisfies a relation

$$\tilde{U} H_{\mathbf{k}_\parallel} \tilde{U}^{-1} = \varepsilon_C H^t_{\mathbf{k}_\parallel}, \quad \tilde{U} \equiv \begin{pmatrix} & Q \\ Q & \end{pmatrix}, \tag{7.4}$$

and $Q = \tilde{P} U$.

The skew symmetric relation is given as

$$\tilde{U} H_{\mathbf{k}_\parallel} = -[\tilde{U} H_{\mathbf{k}_\parallel}]^t, \tag{7.5}$$

R. Takahashi, *Topological States on Interfaces Protected by Symmetry*,
Springer Theses, DOI 10.1007/978-4-431-55534-6_7

namely

$$\tilde{U} H_{\mathbf{k}_\parallel} \tilde{U}^* = -H^t_{\mathbf{k}_\parallel}, \tag{7.6}$$

and the above relation and assumptions lead to

$$\begin{cases} Q H_{\alpha,\mathbf{k}_\parallel} Q^\dagger = \varepsilon_C H^t_{\beta,\mathbf{k}_\parallel}, \\ Q H_{\beta,\mathbf{k}_\parallel} Q^\dagger = \varepsilon_C H^t_{\alpha,\mathbf{k}_\parallel}, \end{cases} \tag{7.7}$$

and

$$\begin{cases} Q V^\dagger_{\mathbf{k}_\parallel} Q^\dagger = \varepsilon_C V^*_{\mathbf{k}_\parallel}, \\ Q V_{\mathbf{k}_\parallel} Q^\dagger = \varepsilon_C V^t_{\mathbf{k}_\parallel}. \end{cases} \tag{7.8}$$

By the above conditions, we have

$$\begin{aligned} \tilde{U} H_{\mathbf{k}_\parallel} \tilde{U}^* &= \begin{pmatrix} Q H_{\alpha,\mathbf{k}_\parallel} Q^* & Q V^\dagger_{\mathbf{k}_\parallel} Q^* \\ Q V_{\mathbf{k}_\parallel} Q^* & Q H_{\beta,\mathbf{k}_\parallel} Q^* \end{pmatrix} = \begin{pmatrix} Q H_{\beta,\mathbf{k}_\parallel} Q^\dagger Q Q^* & Q V^\dagger_{\mathbf{k}_\parallel} Q^\dagger Q Q^* \\ Q V_{\mathbf{k}_\parallel} Q^\dagger Q Q^* & Q H_{\alpha,\mathbf{k}_\parallel} Q^\dagger Q Q^* \end{pmatrix} \\ &= \varepsilon_C \begin{pmatrix} H^t_{\alpha,\mathbf{k}_\parallel} & V^*_{\mathbf{k}_\parallel} \\ V^t_{\mathbf{k}_\parallel} & H^t_{\beta,\mathbf{k}_\parallel} \end{pmatrix} Q Q^*. \end{aligned} \tag{7.9}$$

According to the skew symmetric relation (7.6), the following condition is obtained as

$$\varepsilon_C Q Q^* = \varepsilon_C U U^* = \varepsilon_C \eta_U = -1. \tag{7.10}$$

Thus, systems with the interfacial symmetry are classified as [1]

$$\begin{cases} (\varepsilon_C, \eta_U) = (1, -1), & \text{interfacial time-reversal symmetry (ITRS)}, \\ (\varepsilon_C, \eta_U) = (-1, 1), & \text{interfacial particle-hole symmetry (IPHS)}. \end{cases} \tag{7.11}$$

7.2 Fermi Loop for the IPHS

Here, we assume that H_α is diagonalized by using a unitary matrix $W_{\mathbf{k}_\parallel}$ as

$$W^\dagger_{\mathbf{k}_\parallel} H_{\alpha,\mathbf{k}_\parallel} W_{\mathbf{k}_\parallel} = E_{\alpha,\mathbf{k}_\parallel}, \tag{7.12}$$

where $E_{\alpha,\mathbf{k}_\parallel} = \mathrm{diag}(E_{1,\mathbf{k}_\parallel}, E_{2,\mathbf{k}_\parallel}, \ldots, E_{N,\mathbf{k}_\parallel})$. Correspondingly, by Eq. (7.7), $H_{\beta,\mathbf{k}_\parallel}$ is also diagonalized by $W_{\mathbf{k}_\parallel}$ as

$$(QW_{\mathbf{k}_\parallel})^t H_{\beta,\mathbf{k}_\parallel} Q^* W^*_{\mathbf{k}_\parallel} = \varepsilon_C E_{\alpha,\mathbf{k}_\parallel}. \tag{7.13}$$

By defining $\tilde{w}_{\mathbf{k}_\parallel}$ as

$$\tilde{w}_{\mathbf{k}_\parallel} \equiv \begin{pmatrix} W_{\mathbf{k}_\parallel} & 0 \\ 0 & Q^* W^*_{\mathbf{k}_\parallel} \end{pmatrix}. \tag{7.14}$$

We have

$$\begin{aligned}
\mathscr{K} &\equiv \tilde{w}^\dagger_{\mathbf{k}_\parallel} H_{\mathbf{k}_\parallel} \tilde{w}_{\mathbf{k}_\parallel} \\
&= \begin{pmatrix} W^\dagger_{\mathbf{k}_\parallel} & 0 \\ 0 & W^t_{\mathbf{k}_\parallel} Q^t \end{pmatrix} \begin{pmatrix} H_{\alpha,\mathbf{k}_\parallel} & V_{\mathbf{k}_\parallel} \\ V^\dagger_{\mathbf{k}_\parallel} & H_{\beta,\mathbf{k}_\parallel} \end{pmatrix} \begin{pmatrix} W_{\mathbf{k}_\parallel} & 0 \\ 0 & Q^* W^*_{\mathbf{k}_\parallel} \end{pmatrix} \\
&= \begin{pmatrix} E_{\alpha,\mathbf{k}_\parallel} & W^\dagger_{\mathbf{k}_\parallel} V_{\mathbf{k}_\parallel} Q^* W^*_{\mathbf{k}_\parallel} \\ W^t_{\mathbf{k}_\parallel} Q^t V^\dagger_{\mathbf{k}_\parallel} W_{\mathbf{k}_\parallel} & \varepsilon_C E_{\alpha,\mathbf{k}_\parallel} \end{pmatrix} \\
&= \begin{pmatrix} E_{\alpha,\mathbf{k}_\parallel} & \eta_U W^\dagger_{\mathbf{k}_\parallel} V_{\mathbf{k}_\parallel} Q^\dagger W^*_{\mathbf{k}_\parallel} \\ \eta_U W^t_{\mathbf{k}_\parallel} Q V^\dagger_{\mathbf{k}_\parallel} W_{\mathbf{k}_\parallel} & \varepsilon_C E_{\alpha,\mathbf{k}_\parallel} \end{pmatrix}.
\end{aligned} \tag{7.15}$$

By the skew symmetric relation Eq. (7.6), we have

$$\tilde{U} \tilde{w}_{\mathbf{k}_\parallel} \mathscr{K} \tilde{w}^\dagger_{\mathbf{k}_\parallel} \tilde{U}^* = -\tilde{w}^*_{\mathbf{k}_\parallel} \mathscr{K}^t \tilde{w}^t_{\mathbf{k}_\parallel}, \tag{7.16}$$

$$\to \tilde{w}^t_{\mathbf{k}_\parallel} \tilde{U} \tilde{w}_{\mathbf{k}_\parallel} \mathscr{K} \tilde{w}^\dagger_{\mathbf{k}_\parallel} \tilde{U}^* \tilde{w}^*_{\mathbf{k}_\parallel} = -\mathscr{K}^t, \tag{7.17}$$

and then

$$\begin{aligned}
\tilde{U}' &\equiv \tilde{w}^t_{\mathbf{k}_\parallel} \tilde{U} \tilde{w}_{\mathbf{k}_\parallel} \\
&= \begin{pmatrix} W^t_{\mathbf{k}_\parallel} & 0 \\ 0 & W^\dagger_{\mathbf{k}_\parallel} Q^\dagger \end{pmatrix} \begin{pmatrix} & Q \\ Q & \end{pmatrix} \begin{pmatrix} W_{\mathbf{k}_\parallel} & 0 \\ 0 & Q^* W^*_{\mathbf{k}_\parallel} \end{pmatrix} \\
&= \begin{pmatrix} 0 & W^t_{\mathbf{k}_\parallel} Q Q^* W^*_{\mathbf{k}_\parallel} \\ W^\dagger_{\mathbf{k}_\parallel} W_{\mathbf{k}_\parallel} & 0 \end{pmatrix} = \begin{pmatrix} 0 & \eta_U \\ 1 & 0 \end{pmatrix}.
\end{aligned} \tag{7.18}$$

Therefore,

$$\begin{aligned}
\mathrm{pf}(\tilde{U} H_{\mathbf{k}_\parallel}) &= \mathrm{pf}(\tilde{U} \tilde{w}_{\mathbf{k}_\parallel} \mathscr{K} \tilde{w}^\dagger_{\mathbf{k}_\parallel}) = \mathrm{pf}(\tilde{w}^*_{\mathbf{k}_\parallel} \tilde{U}' \tilde{w}^\dagger_{\mathbf{k}_\parallel} \tilde{w}_{\mathbf{k}_\parallel} \mathscr{K} \tilde{w}^\dagger_{\mathbf{k}_\parallel}) \\
&= \det(\tilde{w}^*_{\mathbf{k}_\parallel}) \mathrm{pf}(\tilde{U}' \mathscr{K}).
\end{aligned} \tag{7.19}$$

Under the transformation, Eq. (7.15), the hybridization is given as

$$V' \equiv W^\dagger_{\mathbf{k}_\parallel} V_{\mathbf{k}_\parallel} Q^\dagger W^*_{\mathbf{k}_\parallel}. \tag{7.20}$$

This satisfies the skew symmetric relation as

$$\begin{aligned} V'^t &= W^\dagger_{\mathbf{k}_\parallel} Q^* V^t_{\mathbf{k}_\parallel} W^*_{\mathbf{k}_\parallel} = \eta_U W^\dagger_{\mathbf{k}_\parallel} Q^* V^t_{\mathbf{k}_\parallel} W^*_{\mathbf{k}_\parallel} \\ &= \eta_U \varepsilon_C W^\dagger_{\mathbf{k}_\parallel} V_{\mathbf{k}_\parallel} Q^\dagger W^*_{\mathbf{k}_\parallel} = \eta_U \varepsilon_C V', \end{aligned} \tag{7.21}$$

where we used Eq. (7.8). Therefore, $\tilde{U}'K$ is given as

$$\begin{aligned} \tilde{U}'K &= \begin{pmatrix} 0 & \eta_U \\ 1 & 0 \end{pmatrix} \begin{pmatrix} E_{\alpha,\mathbf{k}_\parallel} & \eta_U V' \\ \eta_U V'^\dagger & \varepsilon_C E_{\alpha,\mathbf{k}_\parallel} \end{pmatrix} \\ &= \begin{pmatrix} V'^\dagger & \eta_U \varepsilon_C E_{\alpha,\mathbf{k}_\parallel} \\ E_{\alpha,\mathbf{k}_\parallel} & \eta_U V' \end{pmatrix}. \end{aligned} \tag{7.22}$$

Then, for $\eta_U = 1$, the leading-order terms of the Pfaffian expanded in terms of V' are given as

$$\mathrm{pf}(\tilde{U} H_{\mathbf{k}_\parallel}) \sim Z_{\bar{N}} \left[1 + \frac{|V'_{m,m+1,\mathbf{k}_\parallel}|^2}{E_{m,\mathbf{k}_\parallel} E_{m+1,\mathbf{k}_\parallel}} \right], \tag{7.23}$$

where $Z_{\bar{N}} = \prod_{i=1}^{\bar{N}} E_{i,\mathbf{k}_\parallel}$. Thus, Eq. (3.66) is derived.

Reference

1. R. Takahashi, S. Murakami, Phys. Rev. Lett. **113**, 256406 (2014)

Curriculum Vitae

Education

- Postdoc, Department of Applied Physics, University of Tokyo, 4/13–present
- Ph.D., Department of Physics, Tokyo Institute of Technology, 4/10–3/13
- M.A., Department of Physics, Tokyo Institute of Technology, 4/08–3/10
- B.A., Department of Physics, Tokyo Institute of Technology, 4/04–3/08

Research Experience

Postdoc

Advisor: Prof. Naoto Nagaosa, 4/13–present

- Theoretical study of the Berry phase

Ph.D.

Advisor: Prof. Shuichi Murakami, 4/8–3/13

- Theoretical study of topological insulators

R. Takahashi, *Topological States on Interfaces Protected by Symmetry*,
Springer Theses, DOI 10.1007/978-4-431-55534-6

Work Experience

- Postdoc, Department of Applied Physics, University of Tokyo, 4/13–present

Publications

- R. Takahashi, Shuichi Murakami, gGapless interface states between two topological insulators with crystallographic rotational symmetry'h, JPS Conf. Proc. **4**, 011005 (2015).
- R. Takahashi, Shuichi Murakami, "Interfacial Fermi Loops from Interfacial Symmetries", Phys. Rev. Lett. **113**, 256406 (2014).
- R. Takahashi, S. Murakami "Completely flat bands and fully localized states on surfaces of anisotropic diamond-lattice models", Phys. Rev. B **88**, 235303 (2013).
- R. Takahashi, S. Murakami "Thermoelectric transport in topological insulators", Semicond. Sci. Technol. **27** 124005 (2012).
- R. Takahashi, S. Murakami "Gapless Interface States between Topological Insulators with Opposite Dirac Velocities", Phys. Rev. Lett., **107**, 166805 (2011).
- R. Takahashi, S. Murakami "Thermoelectric figure of merit in topological insulators", J. Phys.: Conf. Ser. **302**, 012026 (2011).
- S. Murakami, R. Takahashi, O A Tretiakov, Ar Abanov and Jairo Sinova "Thermoelectric transport of perfectly conducting channels in two- and three-dimensional topological insulators", J. Phys.: Conf. Ser. **334** 012013 (2011).
- R. Takahashi, S. Murakami "Thermoelectric transport in perfectly conducting channels in quantum spin Hall systems", Phys. Rev. B **81**, 161302(R) (2010).

References

Shuichi Murakami

Professor of Condensed Matter Physics
Mailing address: Department of Physics, Tokyo Institute of Technology, 2-12-1 Ookayama, Meguro-ku, Tokyo 152-8551, Japan
Office phone: +81-3-5734-2747
E-mail: murakami@stat.phys.titech.ac.jp

Naoto Nagaosa

Professor of Condensed Matter Physics
Mailing address: Department of Applied Physics, University of Tokyo, 7-3-1 Hongo, Bunkyo-ku, Tokyo 113-8656, Japan
E-mail: nagaosa@ap.t.u-tokyo.ac.jp

MIX
Papier aus verantwortungsvollen Quellen
Paper from responsible sources
FSC® C105338

If you have any concerns about our products,
you can contact us on
ProductSafety@springernature.com

In case Publisher is established outside the EU,
the EU authorized representative is:
Springer Nature Customer Service Center GmbH
Europaplatz 3, 69115 Heidelberg, Germany

Printed by Libri Plureos GmbH
in Hamburg, Germany